KU-316-082

THE ORCHID HUNTER
A YOUNG BOTANIST'S SEARCH FOR HAPPINESS

By Leif Bersweden

THE
ORCHID
HUNTER

A YOUNG BOTANIST'S SEARCH FOR HAPPINESS

LEIF BERSWEDEN

Published in 2017 by Short Books
Unit 316, ScreenWorks,
22 Highbury Grove
London N5 2ER

10 9 8 7 6 5 4 3 2

Copyright © Leif Bersweden

The author has asserted his right under the Copyright, Designs and
Patents Act 1988 to be identified as the author of this work. All rights
reserved. No part of this publication may be reproduced, stored in a re-
trieval system or transmitted in any form, or by any means (electronic,
mechanical, or otherwise) without the prior written permission of both
the copyright owner and the publisher.

Illustrations © Evie Dunne

A CIP catalogue record for this book is available
from the British Library.

ISBN 978-1-78072-334-1

Printed in Great Britain by CPI Group (UK) Ltd,
Croydon, CR0 4YY

Cover illustrations © Evie Dunne
Cover design by Georgia Vaux

Where material has been quoted in this text, every effort has
been made to contact copyright-holders and abide by 'fair use'
guidelines. If you are a copyright-holder and wish to get in
touch, please email info@shortbooks.co.uk

For Mum, Dad, Esther and Naomi

Contents

	Introduction: A Bee's Eye View	11
1.	Shakespeare's Long Purples	23
2.	Stumped by Ireland's Mediterranean Orchid	41
3.	The Loose Flowers of Jersey	61
4.	The Lady of the Woods and Her Hanged Man	71
5.	Orchids of the Emerald Isle	87
6.	Swords of the Hampshire Hangers	105
7.	The Desirable Category of Very Rare Orchids	121
8.	Butterflies and Burnt Tips	143
9.	The Lady's Slipper	159
10.	The Curse of the Coralroot	179
11.	Finding the Fragrants	197
12.	Those Little Green Ones	211
13.	Mimic	227
14.	Queen of the Cotswolds	247
15.	Orchids of the Western Isles	257
16.	Midsummer Musk	273
17.	Holy Helleborines	289
18.	Ghost Hunting	307
19.	August Orchids	319
20.	Spirals by the Sea	337

Introduction
A Bee's Eye View

*'Some have floures wherein is to be seene the shape of
sundry sorts of living creatures, some the shape and
proportion of flies, in others... humble bees.'*

John Gerard, *The Herball* (1597)

Figsbury Ring
June 2001

Down in the hollows, the evening light cast long, exaggerated
shadows. The air was still, the heat of the day lingering on the
breeze. A blackbird sang somewhere in the scrub and around
me grasshoppers chirped: a constant, comforting undertone.
I was crouched in the lower branches of an ancient beech
tree, searching for a route up. In the distance, I heard my
mother call impatiently.

I dropped to the ground and careered through the grass
before heading up the slope towards her. Almost at the top,
I became distracted by something and bent down to take a

11

closer look: swirls of yellow, orange and brown surrounded by three pink petals.

"Mum!" I yelled. A little further on my family stopped and turned. My mother shook her head, before slowly making her way over. As she reached my side, I pointed at my find.

"Mum, this flower looks just like a bee," I said. I was seven years old and had found my first orchid.

My life would never be the same.

Who knows what it is that will spark a lifelong interest in us, rather than just a passing fancy? It seems trite to say that I found that first Bee Orchid beautiful – although I certainly did. It was more the mystery, I suppose, something like a question to be answered – the strange, hybrid 'otherness' of the flower which, even to my naïve boyhood self, signalled a fascinating transgression, a crossing of a line between two worlds that had previously seemed so different.

My interest in the natural world began through my father. As a child, I was fascinated by his hobbies. He was a keen ornithologist, and we would sit for hours in freezing bird hides, binoculars glued to our eyes, snacking on homemade sandwiches. I would bring my notebooks and obediently document the names of the ducks and waders as they flew by. One summer we constructed a sweep net using an old broom handle, a metal frame and some bed linen, and spent hours in the field behind our house capturing insects, taking them home in plastic tubes to identify. These summer outings were the highlight of my childhood, and in those early years, I would follow my father wherever he went.

It was my mother, however, who first introduced me to botany. On our walks she would patiently list the names of plants that I pointed out: red campion, cow parsley, greater stitchwort, meadow buttercup. I bought a notebook and began making lists of flowers. Obsessed with the Bee Orchid, I revisited Figsbury Ring every June, but despite my best efforts I never found it there again.

My family lived in a red-brick vicarage on the outskirts of Winterslow, a sprawling village on the Wiltshire–Hampshire border: a near-perfect environment for a boy who loved plants. In my spare time I roamed the countryside for hours. Equally fascinated by everyday buttercups and rare orchids, I became determined to see as many species as possible, and by the age of twelve I had extensively documented the flora of my local area.

But it wasn't enough. I began making trips further afield to search for rarer species I had singled out from my wild-flower guides. My parents seemed keen to encourage my interest, and at weekends, instead of ferrying me to sporting events, they took me around the countryside in search of these plants.

Not surprisingly, over time my hobby sometimes wore them down. In 2008, on a trip to the New Forest, my father and I went in search of the diminutive Bog Orchid. I had spent several weeks badgering him to take me, promising the trip would be a success. That afternoon I was committed to the hunt, squelching through the peat and recording the different sundews and bladderworts. More than once I fell into the bog, soaking my trousers. After five hours of searching we gave up, and returned disconsolately to the car, where we sat in stony silence. My father was annoyed at having wasted his day off; I was bitterly disappointed. Soon I became aware of the repulsive smell of wet peat seeping from my clothes. My father clearly noticed it too, his jaw muscles twitching as he wound down the window, and we were silent for the entire journey home.

To their enormous credit, my parents continued to humour me, despite a relatively low success rate and substantial petrol costs. Every time we went on holiday, I would insist on making diversions to look for orchids. What I promised would be a ten-minute drive inevitably ended up taking several hours, much to the annoyance of everyone else in the car. My sisters could not understand my parents' indulgence. The three of us would be crammed into the back seat, me reciting the names of plants I saw from the window, while my sisters threw irritated looks in my direction.

As I grew up, my obsession with plants allowed me to bypass many of the trials of adolescence. Instead of facing the terrifying prospect of talking to girls my own age, I spent hours combing the fields and woods for plants. My friends teased me mercilessly; their jokes primarily revolved around me having lewd relationships with flowers.

Ironically, orchids have symbolised romance, sex and seduction throughout history. For centuries, they were believed to have aphrodisiac properties, helping to excite the sexual appetite of both men and women. In 1704, the French writer Louis Liger told the story of Orchis, son of the satyr Patellanus and the nymph Acolasia, who was sentenced to death for laying hands on a priestess at the festival of Bacchus. His father interceded for him and the gods turned him into the flower which still bears his name today. While this story is invariably attributed to Greek or Roman mythology, there is no mention of Orchis, Patellanus or Acolasia prior to the publication of Liger's book, so he probably invented it. In fact, the orchid's name is actually derived from the Greek word *orkhis*, meaning testicle, a reference to the shape of the tubers found in some species.

By the time I finished my GCSEs, I had seen perhaps a third of the fifty or so orchids in Britain. The remainder taunted me from the pages of my wildflower books. Some were completely new to me, while others had simply proven elusive on my excursions. But I knew I needed to see them all.

It was failing to secure a university place at Oxford eighteen months later that gave me the window I needed. As soon as I opened my rejection letter, I knew exactly what I wanted to do. That night I sat my parents down and pitched my gap year project to them. No botanist had ever found every species of orchid native to the British Isles in one season. I intended to be the first.

My parents are both vicars: they are stoic by nature, and little surprises them. And by then, of course, they were accustomed to my peculiarities. But on hearing my proposal they immediately raised a host of practical concerns: how would I fund the project, how would I get about, where would I stay?

From the outset they made it clear they wouldn't be able to cover the costs. I'd worked casually as a gardener around the village for a few years, and with the money I'd saved had purchased an old Vauxhall people carrier – a good choice, as it turned out, as it would hold all of my gear and double as a motorhome if necessary. I would get a job, apply for grants and bursaries, and spend the winter saving and preparing.

Cautiously, they gave me their blessing.

Apart from my family and closest friends, I kept my

gap year plans to myself. My school friends were all scattering: some were off to university, while others were planning hedonistic trips to Thailand, or were joining charitable schemes to build huts in Africa. I was not immune to derision, I knew my orchid obsession was a bit odd in their eyes and I was terrified of ridicule. So when people casually asked about my plans, I lied. A large part of my embarrassment was that I would be travelling around Britain; perhaps it wouldn't have been so bad if I had been jetting off to South America to search for tropical orchids.

Orchidaceae is one of the largest families of flowering plants in the world, second only to the Daisies. There are approximately 25,000 identified species worldwide, a number that increases incrementally each year as more are discovered in the depths of the rainforest. Though individually finicky about their living conditions, orchids generally exhibit high levels of ecological diversity. Many are drought tolerant and have low nutrient requirements, giving them the ability to withstand extreme conditions. Some grow in arid, desert-like environments, in swamps or at the top of mountains, while a few can be found growing as semi-aquatic plants. Even within the Arctic Circle, long after trees and shrubs have deemed the environment too cold, there are orchid species that have made the icy tundra their home.

But even the more favourable environments are not necessarily easy to grow in. More than seventy percent of orchid species grow in the tropics: not on the ground, but on trees. There is barely any competition from ground-dwelling plants in the rainforest canopy, which has given orchids the opportunity to adapt and diversify. This is thought to be one of the reasons why they are so species-rich. Living in the treetops presents significant challenges though: there is very little water, they are exposed to intense UV radiation from the sun and subject to extreme temperatures. Many orchids have adapted to living under these conditions, defying stereotypes and constantly extending the boundaries of where plants can grow. Perhaps this is why they have become so popular.

For most people, the word orchid is synonymous with the exotic, showy specimens available in our supermarkets. Orchids signify wealth and mystery, something precious, rare and fragile from a far-away land. During the nineteenth century, Orchidelirium swept across Britain and Europe as wealthy collectors plundered the rainforests for tropical plants. As the British Empire expanded, opening up previously untouched areas, orchids were shipped back to the UK in their thousands. Plant hunters also brought information about habitats, making orchids easier to cultivate at home. By the turn of the century, hothouses across Europe were filled with vast tropical collections.

I, on the other hand, had no interest in the eye-catching tropical species. As I sat down to plot my quest at the age of eighteen, I needed a definitive list of orchids to find. Estimates of the number of orchids native to Britain and Ireland vary significantly, generally falling between fifty and sixty, meaning there was no obvious finishing line. Having consulted various books, I decided to draw my list from *Orchids of Britain and Ireland* by Anne and Simon Harrap, arguably the most comprehensive, up-to-date guide.

Though the Harraps name fifty-six species, some were problematic and would have to be excised from my list if I were to stand any chance of succeeding. Summer Lady's-tresses, for example, is now considered extinct in Britain. The last reliable sighting of this species was in 1959 in the New Forest, after it fell victim to the effects of land drainage and over-enthusiastic plant collecting.

Migratory species were also troublesome. Orchid seeds are miniscule specks that can be carried vast distances on the wind before raining down on Britain. Very few species actually survive here though, as requirements for growth are typically highly specific. Those orchids that manage to establish themselves are often treated with suspicion by experts, given the unscrupulous practice of some botanists of introducing plants deliberately. It is difficult to prove natural occurrence, so I decided not to include continental species on my list. From the Harraps' guide this meant leaving out the recent arrivals of Small-flowered Tongue Orchid in Cornwall and Greater Tongue Orchid in the Channel Islands.

The final and perhaps most controversial exclusion was the Ghost Orchid. Living up to its name, this little plant has always proven elusive in the UK, but recently we have come very close to losing it altogether. So close, in fact, that it has already been declared extinct once, only for it to be rediscovered in Herefordshire during the summer of 2009. There have been no confirmed sightings since, meaning the chances of me finding the Ghost were next to none. For this reason, I decided to leave it off my list.

And if I did manage to find one? It would be the most exciting day of my life.

This left me with a list of fifty-two species. Finding these orchids in one summer would be a superhuman endeavour. The orchid flowering season in Britain is roughly six months long, beginning in April and stretching through the summer to late September. Different species flower at different times, so my journey would be a squiggle across the map. I would have to go up and down the country, constantly retracing my steps. The length of the flowering period varies between species too. Most orchids flower for only a week, while some barely manage two or three days. Such is the precision of these flowering windows that in many cases I would have only one opportunity to find a particular species. And if I missed even one, my project would fail.

Timing would prove crucial, not least because the variability of British weather conditions would inevitably have an impact on flowering periods. I would need to be flexible and spontaneous, ready to abandon my carefully laid plans in an instant.

Over the course of that winter I waited and fretted. I got a job stacking shelves in Waitrose and spent my spare time applying for bursaries and grants, as well as reapplying to study biology at Oxford. Weather was a constant source of worry. If the winter was too warm, the orchids would flower early and I would be likely to miss them. If the weather was cold for too long, it could reduce the numbers that flowered. I needed advice from local experts to keep me informed on the progress of local populations, so I haunted the internet forums popular with botanists. And I would need to rely heavily on my car to reach far-flung corners of the country at short notice – an investment that had so far proved disastrous. After extensive repairs, the Vauxhall was still giving me trouble and spent more time in the garage than out.

Despite all this, I was excited beyond belief. As I stacked shelves with cartons of juice, I daydreamed of my orchid summer: of months spent exploring moors, copses, limestone escarpments and bogs, climbing mountains, visiting remote islands and scouring industrial estates. I didn't know it then, but there would be heartbreak, loss and more than my fair share of technological disasters. If I succeeded, I felt I might finally satisfy my own obsession with orchids, but also understand the depth of our long-held national obsession with them: their ability to capture the hearts of so many, to drive people crazy, incite terrible crimes of passion and bring out our very best and worst.

1

Shakespeare's Long Purples

'There with fantastic garlands did she come
Of crow-flowers, nettles, daisies, and long purples:
That liberal shepherds give a grosser name,
But our cold maids do dead men's fingers call them.'

William Shakespeare, *Hamlet* (circa 1601)

Dorset
May 2013

Orchids are the most charismatic flowers in the Plant Kingdom. Each species has its own personality: there are cheats and thieves, court jesters and hipsters, psychics, librarians and royalty. And each flower, in turn, comes with its own idiosyncrasies.

They also assume a multitude of colours, shapes and sizes. In Britain there are orchids in the shape of spiders, lizards, butterflies and monkeys; there are even some that play dead. Some are only a few centimetres tall, while others can reach over a metre. Ask an orchid enthusiast about their interest and it won't be long before they are showing you albums dedicated to pyramids, slippers, soldiers and birds' nests. Each species smells and feels different too. Some are noisome, such as the Lizard Orchid, which gives off a revolting stench of billy goat; others are delicately perfumed. Some are furry and produce scents so subtle they are undetectable by humans.

The winter of my gap year seemed to stretch on for ever, cold and miserable. My vision of walking through swathes of wildflowers in the sun was quickly becoming hidden behind the rain-lashed windows of a never-ending winter. As I tidied supermarket shelves, I wondered, not for the first time, whether spring would ever come.

February rolled into March and then April, and still the clouds hung low. I spent days meticulously planning and replanning my summer, distracting myself by tracking down sites for rare species and flicking repeatedly through the well-thumbed pages of my orchid books.

Winter is always a frustrating time for botanists. The season has a lot to offer ornithologists and other naturalists, but there is little for the plant lover. Only the truly enthusiastic turn to tiny green mosses or leafless trees to satisfy their botanical needs; in the absence of flowers, many of us go into hiding.

As the end of April approached, the anticipation of seeing my first orchid of the season gripped me and wouldn't let go. The air was warmer and with agonising slowness, the tell-tale signs of spring began to appear. Fresh leaves emerged from the hawthorn, the dawn chorus resumed and the air was filled with the scent of daffodils, crocuses and freshly mown grass.

One day after work, I went for a walk in Bentley Wood,

intending to gauge the progress of the transitioning season. Bentley is just down the road from my village, and is more famous for butterflies than orchids. As a child, I searched with my father for pearl-bordered fritillaries and white admirals along the woodland rides there each summer. One August, we waited for hours with scores of butterfly enthusiasts to catch a glimpse of the purple emperor, and were rewarded handsomely when one flew down and landed in the boot of our car. Its wings shimmered in the sunshine, opening and closing with audible, sharp clicks.

But at winter's end, Bentley Wood is a very different place. It had been another miserable, wet April day. The damp smell of sodden leaf mulch lay heavily in the air and rivulets of water trickled past me as I climbed the twisted path up the hill.

An ancient bank of earth ran along the edge of the wood; it once would have served to keep livestock out. In Medieval times, woodland was bounded by banks and ditches to demark ownership or parish limits. These perimeter boundaries have long since succumbed to nature and become woven into the fabric of the wood, but they remain a reliable indicator of ancient woodland.

I followed the boundary south, alternating between the path and a bank covered in wood melick and sedges, and wading through ankle-deep dog's-mercury. Two jays erupted into argument, startling me. One of them flew down and landed in a white-starred blackthorn on the woodland edge, expertly avoiding the dark thorns protruding from its depths. It caught my eye and leapt upwards with a shriek, snagging a wing on a cluster of thorns and surrendering a single bright-azure feather.

The beech trees here were old and gnarled, their trunks a pallet of merging greys. Desiccated boughs from a forgotten storm lay scattered among knotted roots thicker than my arm.

Caught up in admiring my surroundings, I almost missed the rosette of brown-splotched leaves nestled behind a tree stump. These were the leaves of the Early Purple Orchid, one of the earliest orchids to flower and a likely contender for my first species of the year. I was pleased with the find, but to my dismay, there was no sign of a developing flower – confirmation that the season was going to be severely delayed. This would make it more difficult to predict when different species would flower.

Leaves are the first tangible stage of an orchid's life cycle. Some produce basal rosettes where the foliage grows close to the ground, while others produce leaves on the stem as they grow. As the season progresses, the stem, or flowering spike, grows a head of flowers called an inflorescence. Orchid flowers are made up of six petal-like structures: three on the outside called sepals and three on the inside called petals, one of which looks different from the others and is called the lip. The flowers are pollinated, often by insects, producing tiny, dust-like seeds so light that a small breath of wind is enough to disperse them.

Orchid seeds are special. Most seeds are sent off into the world by their parent plant with a packed lunch of starch and other sugars. This food store, known as the endosperm, is meant to tide over the young seedling until it is able to begin photosynthesising of its own accord. This allows the seed to germinate quickly under suitable conditions. Orchid seeds, however, don't have an endosperm. As a result, they are miniscule, typically weighing two to eight micrograms – less than one-fifth of the average grain of sand. The early growth of an orchid is extremely slow: nothing resembling a normal seedling is visible for months or even years, as most species will not flower within the first four or five years of their life-time.

How, then, does the young orchid survive its early life without an endosperm? The answer: with the help of a fungus that grows in its roots. More than 90 percent of the world's plants are thought to be associated with fungi at some point in their life cycle. Orchids take this relationship to the extreme and are entirely dependent on them while they are under-ground. Once they have germinated, most species maintain a functioning relationship with their fungal partner, which has allowed orchids to adapt and prosper in poorer habitats like deeply shaded woods, or the nutrient-poor soils of heaths and bogs. This relationship between orchids and fungi is known as a mycorrhiza (literally 'fungus root').

That day in Bentley Wood, I found four Early Purple rosettes altogether. It was impossible to tell how old the plants in front of me were, or whether they would flower that year. But it was a start, the first sign of many more to come.

When May arrived, a string of fine days brought the first butterflies of the year. Orange-tips and brimstones danced around the garden, waiting patiently as the trees burst into leaf. In the woods, exposed pathways were suddenly enclosed in a protective shroud of green. That year, an extremely hot May bank holiday gripped the nation. The headlines read 'All

Hail the Arrival of Summer' and those in the botanical world began busily estimating the length of the delay to the season.

I had my own concerns. The excitement over the sudden influx of summer made me wonder whether the year's first orchids would come and go too swiftly, as plants struggled to reorient themselves around the season. Once again, my car was in the garage. We were closing in on the middle of May and I had yet to see one orchid. My schedule had already been blown; I carefully recalculated my plans and waited for my car's return, as stories of flowering orchids across the south of England started coming in thick and fast.

I relied heavily on online nature forums – where experts and enthusiasts post photos and sightings from their local areas – for the latest news on flowering times. Given how distorted the season was set to be after the long winter, this resource would prove increasingly invaluable throughout the summer.

Even my parents were worried, checking with me constantly to make sure I wasn't going to miss the first species. On top of this, I had only passed my driving test four months previously, and while I had been driving to and from work since, I was about to embark on my first long-distance journey. Concerns about my safety, whether the satnav would work and what would happen if I put the wrong fuel in my car were expressed repeatedly in the Bersweden household in the days leading up to my first excursion.

The forums had exploded with sightings of Early Spider Orchids, one of Britain's rarer species, on the south coast and made for torturous reading while the delays continued with my car. Already, my head was filled with fear of failure.

After a particularly agonising weekend in mid-May,

I finally set off for the Dorset coast to hunt for the Early Spiders. I passed through Wareham and twisted my way towards Swanage. Corfe Castle came and went, giving way to heathland bathed in yellow flowering gorse, and then the brightly coloured cottages of the Swanage sea-front.

Just outside town, on the Jurassic cliffs, lies Durlston Country Park, a nature reserve that has become one of my favourite botanical haunts. Purbeck limestone has been quarried along this stretch of coastline since the first century AD and transported for use around the country, most notably to rebuild parts of London after the Great Fire of 1666. Old coastal quarries west of Durlston at Winspit, Seacombe and Dancing Ledge provide unique habitats that have been left undisturbed for decades; these have been exploited by Early Spider Orchids.

I pulled into the car park at Durlston, and sat there, taking a moment to collect myself. This was it: the months of waiting were finally over and my adventure was about to begin. There were fifty-two orchids out there waiting to be found. Eventually, booted up and suitably camera-laden, I set off through the scrub and out on to the hillside.

The wind nearly blew me off my feet. Wiping my bleary eyes, I held a gate open for an elderly couple who mouthed their thanks, not even bothering to shout. But as I battled my way along to the first meadow, I reached a small copse and felt the wind drop dramatically. Immediately, I could hear again. The wonderful melody of a skylark floated down from above and a blackbird plinked from a nearby blackthorn. My eyes settled on the thousands of cowslips that formed a sea of yellow stretching all the way up the meadow and it was as if a switch had been thrown somewhere deep inside me: I was in plant-hunting mode. One minute I was fighting my way across the downs resenting the weather, the next I was striding along completely impervious, my eyes sweeping the grass either side of the path like a search light.

And then it happened: orchid number one appeared in front of me several feet from the footpath, not an Early Spider, but an Early Purple.

I stopped short, surprised by how quickly I'd found my first species. There were the blotched leaves I'd seen a few weeks before, a darkly stained stem and a loosely arranged inflorescence of deep-pink flowers. Amusingly, each flower holds two sepals aloft, giving the illusion of tufted rabbit ears.

In a quick search of the area, I counted roughly thirty Early Purples. There was quite a bit of variation in colour even in this small colony, from a rich violet through to pastel pink; a few were so pale they were almost white.

The Early Purple is one of the most widespread species of orchid in Britain. It is a delight to find, whether among carpets of bluebells in an ancient copse or growing with primroses and cowslips in an old hay meadow. The first British record of a sighting was made by William Turner in 1562. In his *Herball*, Turner writes, 'there are divers kindes of orchis... one kinde... hath many spottes in the leafe and is called adder grasse in Northumberland.'

Few of our native orchids have ever been common enough to warrant the assignment of local names. The Early Purple has almost a hundred. The English botanist Nicholas Culpeper wrote in *The English Physitian* (1652) that it 'hath gotten almost as many several names attributed to the several sorts of it, as would almost fill a Sheet of Paper; as Dog-stones, Goat-stones, Fool-stones, Fox-stones, Satirion, Cullians, together with many others, too tedious to rehearse'.

While many of these names have fallen out of use, some have survived the test of time. In Cheshire, for example, it is called Gethsemane, while in Somerset it is known as Adder's Flower. Elsewhere in the country local botanists may recall names like Cuckoos, Regals, Bloody Man's Fingers, Soldier's Jackets, Kettle-cases and Dead-men's-fingers. The latter epithet has led many to believe that Shakespeare's 'long purples' are in fact Early Purple Orchids.

The Latin name for this species, *Orchis mascula*, makes reference to the robust testicular form of the plant's tubers, storage organs on the roots that contains sugars produced during photosynthesis. The generic name *Orchis* literally means 'testis'; however these paired tubers are not only found in Orchidaceae. For example, the Solanaceae family produce tubers that we are all familiar with: potatoes.

In *Orchis*, one of the two tubers will have survived from the previous winter, while the other is forming ready to sustain the plant during the next one. Before the arrival of tea and coffee in the 1700s, tubers from *Orchis mascula* were used to make a drink called salep, consumed across Europe from Turkey to Britain. Ground tubers were added to water which would then be sweetened and flavoured. The species is now endangered in Turkey, where the drink is still popular.

Beneath its beautiful exterior, the Early Purple Orchid is actually a dastardly trickster. The centre of the flower, where the pink fades into white, sports a set of small purple spots. These are thought to guide visiting insects towards the mouth of the spur, the structure containing nectar. And here's where the deceit lies. The Early Purple Orchid doesn't produce nectar. It cheats its pollinators into thinking they're going to get a sweet, sugary meal and in doing so ensures its pollen gets stuck to the insect's head.

An orchid's pollen is attached to sticky structures called pollinia – thin stalks with fuzzy heads, which look a bit like miniature microphones. When an insect enters the flower, the pollinia adhere to its head and stand vertically, almost like small antennae. As the insect flies away an extraordinary change takes place: the stalks bend ninety degrees and lean forward, adopting a horizontal position. It is crucial that this occurs after the insect has left the first orchid, but before it arrives at the next one, because once the pollinia have adopted this stance they will come into contact with the stigma of the flower the insect visits first, which initiates pollination.

I finished photographing the Early Purples and collected my gear before rejoining the path. One orchid down, but I had yet to find the one I'd come in search of. I continued on made my way along the coast, walking beside a weather-beaten stone wall, burdened with ivy that bordered the edge of the field.

I strolled along, listening to unseen chiffchaffs and totting up a list of wildflowers in my notebook. Yellow cowslips, crosswort and the first bulbous buttercups of the year were followed by common vetch, salad burnet and hairy violets in various shades of pink, red and purple. All around me nature was busy: wrens whirred, bees bumbled and ferns unfurled.

The Early Spider Orchid, or *Ophrys sphegodes*, grows in very few places in the UK, only in scattered locations along the coast of Kent, Sussex and Dorset. Here, though, it often appears in the thousands. First recorded in 1650 by a Dr Bowle in Northamptonshire, this species has been one of the most adversely affected by agricultural intensification. By the

1800s, inland locations for this little orchid were already rare, with many populations extinct by the turn of the nineteenth century. Writing in 1950, J. E. Lousley laments the decline of *sphegodes*, conceding that 'it is fairly safe to say that the last known localities were merely those which had dodged the plough longest'. As a consequence, the Early Spider has been lost from 73 per cent of its historical range and is now a Red Data Book species protected under Schedule 8 of the Wildlife and Countryside Act (1981).

But not all change is bad. During work on the Channel Tunnel in Kent in the 1990s, the decision was made to place some five million cubic metres of chalk material excavated from under the sea at the foot of the cliffs around the entrance to the tunnel. Over the following years, the site was landscaped and opened as a nature reserve called Samphire Hoe. In 1998, Early Spider Orchids were discovered growing at the site, thought to have arrived by seed from the small population living on top of the cliffs. There are now an estimated 10,000 plants established on that short stretch of Kentish coastline, testimony to the opportunistic nature of orchids when suddenly presented with appropriate conditions – in this case the chalk.

Heading down into the scrub that ran along the top of the cliffs, I explored the grassy areas among the bushes. The small blue buds of chalk milkwort were swelling in clusters among the lemony bottle-brush flowers of fresh spring sedge. In one little clearing the bracken had begun to take over: everywhere I looked there were coiled croziers and fronds unfurling. In another, there were more Early Purple Orchids. The wind was almost non-existent here and I was hit by the glorious coconut fragrance of the gorse's golden flowers. In an instant, I was eight years old again, clambering over rocks on the Devon coast path with my cousins.

Intent on finding the Early Spider, I emerged from the gorsy hollow and back out into the open. The path wove through a series of small hillocks, the hollows in between providing a safe sanctuary from the strong prevailing winds. The grass here was much shorter, nibbled to within an inch of its life by rabbits. Glancing to my left, I picked out the miniature profile of the species I'd come in search of. My glee at seeing a rare Early Spider Orchid made me grin from ear to ear; I even did a little dance.

I took out my camera and began taking photos. Early Spiders are exquisite little plants. Standing roughly six centimetres tall, the stem, sepals and petals are a vivid lime green which makes them easy to spot in the short turf. The central petal, or lip, of the flower is an exception. Supposedly resembling the legless body of a large spider, it is bigger than the other petals and a dark chocolatey brown. Scrawled across the middle is a pale, silvery-blue letter 'H' or 'π'. As I looked closer, I noticed two round eyes gazing up at me and hairy haunches bristling in the wind. The Early Spider Orchid's Latin name, *Ophrys sphegodes*, pays tribute to its 'wasp-like' nature. Former names like *aranifera* (Greek for 'spider') also relate to its appearance. Despite seeming spider-like to human eyes, it has evolved to attract the male solitary bee *Andrena nigroaenea*.

Early Spider Orchids are one of the four species of the genus *Ophrys* that can be regularly found growing in Britain, the others being Bee, Fly, Early Spider and Late Spider. Their flowers are remarkably insect-like and have a fascinating, yet diabolical sex life. While most plants attract pollinators with the promise of nectar, these orchids lure them in with the promise of bee sex. This deception is accomplished by imitating the scent, appearance and texture of virgin female bees.

A passing male bee senses the orchid from afar, drawn by the alluring female smells released by the plant. He's delighted to find what he perceives to be a female resting on the flower, her head buried among the petals. Excited by this, he alights and attempts to mate with the 'female', often vigorously and for prolonged periods. During these fruitless exertions, the bee knocks into the column – the reproductive structure consisting of both male and female parts – which drops two tiny, sticky pollen sacs onto the bee's back. Eventually, the bee gets frustrated by the lack of action and buzzes off in search of a more enthusiastic partner.

The sexual frustration of the bee is essential to the orchid's scheme. In an ideal world, the pollen from one plant will be deposited on another to ensure a mixing of genes. Determined to avoid repeating his mistake, our male bee flies far away from the orchid and its neighbours. However, upon encountering another population, he immediately falls for the ruse all over again, this time depositing the pollen sacs on the new plant, elegantly ensuring sexual reproduction. The new orchid will smell, look and feel ever so slightly different. Not only does this make the bee think he'll have more luck this time, but it also smartly prevents the bee from learning not to visit the orchid again.

Each one of these insect-mimicking orchid species has fine-tuned itself over evolutionary time so that it smells, looks and feels exactly like the female of a single pollinator species. It is an extraordinary fraud, entirely masterminded by a plant.

I got down on my stomach and lay still, eyes focused on

the tiny orchids in front of me. With each gust of wind, the plants shivered and seemed to draw their spiders closer, clustering for warmth. Each one gazed into the distance and waited patiently for a passing *Andrena* bee. Wishful thinking, I thought, the afternoon too cold and blustery for insects.

I continued on my way, filled with a bouncy optimism and stopping in almost every little hollow to admire the little spiders quivering in the wind. By now I had reached the end of the nature reserve and was following the cliffs west towards Dancing Ledge. The coast path undulated, sloping down to accommodate small beaches and pebbled shores. Gulls squabbled in the air above me as I passed cascades of cowslips splashed with the pink of Early Purple Orchids.

Dancing Ledge is a small, disused limestone quarry at the base of the cliffs just south of Langton Matravers. Access to the ledge involves a short scramble down a steep path. The cliffs here are striped with beds of rock so even they look almost ruler-drawn. Here I found more Early Spider Orchids dotted in the turf like little green beacons.

Glancing around, a spike of pink caught my eye. Nestled in the grass were three orchids similar to the Early Purples I had seen earlier. These were different, though. They were stunted and the flowers were hooded by sepals with distinct parallel veins the dull green of an old copper coin. The petals, too, were darker in colour than the Early Purples, and the lip frilled around the edges. The leaves, which in *Orchis mascula* almost always have some spots, were completely unmarked.

These were Green-winged Orchids, *Anacamptis morio*.

My only past experience of the Green-winged Orchid had been several years earlier when I was fifteen. My father had just presided over the funeral of a man who had devoted his life to nurturing a colony of these orchids on his back lawn. Plans by the new owners to build a swimming pool in the garden meant the orchids were in grave danger. I had been invited by his widow to dig up as many plants as I wanted if I thought it would be possible to save them. They were, she said, his legacy.

A few days later, my father and I drove over to her house, equipped with spades and a large groundsheet, and dug up two square metres of turf full of Green-winged Orchids. My father suggested we transfer them to a corner of the church-yard. It was twilight by the time we arrived. Had anyone been passing the church at this late hour, they would have seen the vicar and his son lift a heavy tarpaulin out of the car and proceed to drag it across the graveyard, before digging in a discreet corner, then carefully lowering the tarpaulin into the ground. Imagine the gossip in the parish the next day.

The graveyard orchids flowered for the first few years, but have since dwindled in number, too disturbed by the change in conditions. But it was worth trying to save them. The Green-winged Orchid is declining with dramatic speed. Once a widespread, often common, component of hay meadows, it now only occurs in small isolated populations. Another victim of the plough, fertilisers and the decline of traditional hay meadow farming.

It is linked to the Early Purple by more than just appear-ance. In the sixteenth century, it was thought to be the female of the Early Purple. It was known as the Fool Stone and thought to be female because the 'stones', or tubers, are smaller than in the Early Purple Orchid, or Male Fool Stone. To this day, Green-winged and Early Purple Orchids share numerous local names, such as Bloody Man's Finger, and

Cuckoo-flower. Here in Dorset they are both known as Gid-dyganders.

I sat down, shielding myself against the wind and the first few spots of rain that swept in from the sea. In my first afternoon of hunting, I had seen the first three orchids of the year. They had been straightforward sightings, all orchids I had found in previous years, but nonetheless I felt delighted.

The species on my list that I had never seen before would no doubt prove a greater challenge.

2

Stumped by Ireland's Mediterranean Orchid

'There must be grazing overhead, hazel thickets,
Pavements the rain is dissolving, springs and graves,
Darkness above the darkness of the seepage of souls
And hedges where goosegrass spills its creamy stars.'

Michael Longley, *In Aillwee Cave* (1991)

Ireland
May 2013

Bleak, cold and barren are probably not adjectives one would associate with the environments typically graced by orchids. Our image of them tends to lead us rather to hothouses or tropical rainforests, something compounded by the numerous stories of Victorian plant hunters and their adventures in the jungle. The west coast of Ireland is far from our expectations: rugged, austere and, more often than not, rainy.

The success at Durlston, now several days behind me,

brought renewed confidence as I finalised my plans for the week ahead. I was quickly realising that the dynamic nature of the flowering season would put my carefully laid plans out of kilter. A certain amount of adaptability would be required in order to find the remaining forty-nine species. My next planned target had been the Lady Orchid, but the late season had heavily delayed the plants down in Kent. It would be another week before they started to flower. Instead, I decided to head west, to Ireland, to hunt down the two May orchids that can't be found in Britain: Dense-flowered and Irish Marsh Orchids.

Over the winter, I'd spent a long time discussing my plans with my godfather, Michael. He is a kind-hearted, spirited man with a laugh that echoes around the room long after he has left. A loudly spoken Geordie, forever amused by life's great ironies, Michael was a Roman Catholic priest for forty years in Yorkshire. When not presiding, he is invariably dressed in a brightly coloured cashmere jumper and beige corduroys, and can be found smuggling mint imperials between meals despite the warnings from his doctor. Seeing this as a perfect opportunity to spend time together, we hatched a plan to visit Ireland in the spring.

Michael met me at Shannon airport just outside Limerick. I'd flown from Gatwick on a crammed Ryanair jet absorbed in my book about the Burren.

We drove for an hour and a half, eventually leaving the motorway and well-manicured A-roads for the rough country lanes that took us towards the small town of Ballybunion, famous for its golf courses and wide sandy beaches. We passed a series of chalked slates advertising small bags of salted periwinkles for sale and hot seaweed baths. In the centre of town stood a statue of Bill Clinton playing golf.

We bounced up a pot-holed hillside and ground to a halt in the grass outside a boxy, weather-beaten cottage. Its thick stone walls were pale peach, the paintwork chipped and peeling. Net curtains hung limply in the grubby windows and the slanting roof was simply a large sheet of corrugated iron. I was not convinced. But like a perfect metaphor, the inside was an entirely different story: a cosily lit living room, the air warm from the peat fire which was flickering softly in the burner. Pale pine panels striped the walls. After the long journey, it was sweet relief to slump into one of the squashy sofas.

My first target was the Irish Marsh Orchid, one of the species I'd never seen before. I'd spent some time on the plane trying to memorise the intricate pattern of wiggles and squiggles found on the flower, as well as various other identifiable characteristics. I wanted to be sure I knew what it was when I found one. The marsh orchids are notoriously difficult to identify. Not only are they very similar, but also extremely variable both within and between populations. On top of this, the different species will readily cross with one another, producing a series of hybrid offspring with intermediate characters. In other words, the marsh orchids are still rapidly evolving.

The west coast of Ireland is an indomitable, stalwart land-scape stretching from the county of Donegal in the north, down through Sligo, Mayo, Galway, Limerick and Clare, eventually reaching the rain-lashed shores of Kerry, where we were. I got the impression that the lush green pastures hadn't changed for centuries. This is a corner of Ireland infused with wildness, and life here moves at a slow, rhythmic pace.

My first few days were spent exploring the quiet towns and tiny hamlets dotted along the wind-whipped coast. We visited the Dingle Peninsula, a ragged thumb of land that protrudes into the North Atlantic whose tip is the mainland's most westerly point. Dominated by the dark peaks of Mount Brandon, this beautiful, jagged landscape is a hotchpotch of early Christian chapels, holy wells, idyllic villages, prehistoric ring forts and ancient beehive huts, all woven together by an intricate network of narrow country lanes.

The cliffs, battered by the wind and waves were slowly crumbling into the ocean. Sheep, painted pink and turquoise, lurking around every corner, sheltered under the rocky ledges by the side of the road. Ahead of us emerald fields, glistening with recent rain, within an irregular mesh of dry stone walls, swept down to the sea.

We stopped for a stroll and some tea and banana cake in the pretty town of Dingle. Too early for the school holidays, it was fairly quiet and we quickly adopted this routine in each town we passed through. I jigged down the street to the sound of traditional Irish music, much to Michael's amusement, past tiny idiosyncratic shops filled with local pottery, honey and herbal tea.

Just like my parents, Michael loved to visit churches – provided they were a long way from his former parishes – and so we paused briefly in Dingle's St Mary's church. When I was a child, our family holidays to the Lake District, Yorkshire and Northumberland were never complete without visiting a long line of local churches. Just as I drew up lists of rare plants to go and see, my parents would earmark parishes with famous, tumbledown chapels and age-old oratories. While my parents prayed, my sisters and I obeyed the unwritten rule of stoic, sombre silence. We wandered slowly between pews and pillars, browsing last year's Christmas cards still on sale. My favourite part of these visits was the stained-glass windows: I find the glint and twinkle of the multicoloured panes of glass strangely hypnotic. One window in a dilapidated church hidden in the folds of the Yorkshire Dales has an orchid, the mythical Lady's Slipper, glimmering gold and burgundy.

Back on the road, Michael and I drove right to the end of the peninsula. The sun had firmly established itself and we pulled over to take photos of the surrounding scenery: an incredibly blue sky above a glittering Atlantic Ocean, isolated islands and crumbling cliffs.

At one point the road verged on a small marshy area and my eyes were drawn to a plant with an inflorescence packed with small pink flowers growing beside a mirror-like pool of water. It was an orchid. An Irish Marsh Orchid? It couldn't be; it was the middle of May and according to my books that would be early in a normal season, let alone a late one.

I jumped down from the roadside and landed among the rushes. The plant was still tightly in bud with the exception of a single floret. Bending down, I lifted the lip of the flower with my finger in order to inspect the loopy patterns so characteristic of the genus *Dactylorhiza*. It was certainly a marsh orchid, and my fourth species of the year, but I couldn't tell which one. Of all our native species, the dactyl – or marsh – orchids, named after the finger-like projections in their root system, are among the most difficult to identify. Not only is each species extremely variable, but they have an annoying tendency to hybridise given the smallest opportunity. The resulting plethora of intermediate hybrids makes it so difficult to tell parent from child that many botanists around the country have completely given up on them.

The plant I had on the Dingle Peninsula was particularly puzzling. It didn't help that it was the first one I'd seen since the previous summer. I consulted my go-to guide on British orchids, by Anne and Simon Harrap, and immediately found a good place to start. Ireland is host to only four species of marsh orchid: Northern, Pugsley's, Early and Irish. It was far too early in the season for the first pair so I was already down to two species. Reading the descriptions in Harrap, it seemed relatively easy to distinguish them. Irish Marsh is generally darker, with a broad three-lobed lip patterned with deep-purple double loops; Early Marsh Orchids usually have paler flowers and a very pinched lip. The key difference, it seemed, was that the Early Marsh held its sepals aloft, like a celebratory sports fan, rather than horizontally. I glanced down at the sepals which were held at 45 degrees: not quite celebratory but not exactly unhappy either. Typical.

No matter how hard I tried, I couldn't make my mind up. I took as many photos as possible, focusing on the intricate spotted markings on the lip, and sent them to Steve Povey, a fellow botanist and orchid specialist. I was probably just being a bit slow after months out of practice, but I felt sure Steve could help me out.

Satisfied with the sighting, although frustrated at my rusty identification skills, we set off again around the peninsula, stopping only to visit the 2000-year-old beehive hut settlements – little igloos of stone dotted across the hillside.

The next couple of days were spent in and out of the cottage, alternating between deep conversation and contented silence. Michael was curious about what it had been like for me growing up with two vicars for parents. It was certainly an unusual situation. My parents were not only involved in the church, but in my schools too, so all my friends knew who they were. My father would come in once a week to give assemblies at my primary school and I would dutifully listen to every word he had to say as if he were no different from all the other teachers. As I got older, it became harder. During my entire secondary education, my mother was the school chaplain. Around my GCSEs, I went through a period of intense mortification whenever she stood up to give a talk to the whole school.

To my parents' immense credit, they've always assured me I could make my own decision about whether to share their beliefs. They were never pushy, and never tried to make me believe anything I didn't want to. Instead, I've grown up with all the support I needed to pursue my own interests. Michael, too, had always shared this attitude and, rather than plying me with religious gifts, had always taken great pleasure in buying me the latest Harry Potter book. I let on less than he'd have liked about my personal beliefs in our conversations, though; I was still finding my feet in a world where sharing came less than naturally.

Michael was in a jovial mood as we set off on Saturday morning to tour the Ring of Kerry, the next peninsula south of Dingle, as finger-like as the roots of the dactylorchids. The views from the mountains here were breath-taking; I didn't know where to look. The roadside was lined with Irish spurge, St Patrick's cabbage and the occasional Early Purple Orchid on the old stone walls. Yellowing tussocks of grass grew within the birch and alder carr that lined the loughs at the bottom of the valley, and I was so busy admiring everything from the window that, for the first time I can remember, I completely forgot about taking photographs.

When we returned to Ballybunion, I took advantage of a beautiful evening to walk up the road and onto the moor at the top of the hill. It wasn't long before I was exploring the high grassy banks on either side of me; dog-violets were abundant but not much else was in flower. A cuckoo called in the distance, the first of the year. A gentle breeze swept across the moorland, rustling the brown skeletons of last year's bracken. Pipits skipped from post to post, always staying a few metres ahead of me. After a few minutes, I found a hole in the fence. It gaped invitingly, and boggy pools full of sedges enticed me to enter. I glanced back down the lane, then slipped through into the field. And there right in front of me was another marsh orchid, growing at a bizarre angle halfway up the vertical bank.

Fantastic, I thought, as I squelched my way over, this time I would be able to confidently identify my fourth species of the year. My first thought was how different it looked from the one I had found earlier in the week. This orchid had paler, pinker flowers and the lip was more pinched. Disappointment and frustration met my excitement head on. The flowers were too pale for Irish Marsh Orchid and yet they had rich purple tinsel trails on the lip and a few small spots on the leaves. Their sepals were almost horizontal. A local variety of Irish Marsh perhaps? I couldn't be sure. Further exploration resulted in two more plants, both with the same pallid flowers and heavy markings. Once more I had failed to identify a species in situ, and would have to rely on a more expert opinion.

On returning to the cottage, I found a reply from Steve about my marsh orchid on the Dingle Peninsula. His puzzled comments made me feel better. Unable to come to any conclusion, he offered to send my photos to orchid geneticists at Kew Gardens. Jumping at this fantastic opportunity, I almost missed his parting comment: 'the *praetermissa*-type lip is most interesting, considering that species is not found in Ireland.' *Praetermissa* refers to the Southern Marsh Orchid, a species I had originally dismissed upon reading that it wasn't found in Ireland.

A sudden excitement gripped me. Could I really have just discovered a Southern Marsh Orchid in Ireland? Then another thought quickly brought me back down to earth. This couldn't be a Southern Marsh Orchid, a species which would ordinarily flower in early June; this plant was coming into flower in mid-May during a spring that had only just got started. And yet... the shape of the lip, the colour of the flowers and the vague fine-spotted markings all looked like *praetermissa*.

When I told Michael this, he wisely remarked: 'We are so often reminded that we know so little of life.' There was always a chance this could be a new species for Ireland. I replied to Steve with a simple message and attached the photos of the other marsh orchids I had found on the hillside, in the hope that he could shed some light on these as well.

The Burren rises, lunar and desolate, in the north of County Clare. On the surface it appears to be nothing more than a barren sea of pale limestone, banded with shale and clay, earning it the Gaelic name *boireann*, meaning 'rocky place'. During the last ice age, monstrous glaciers scraped away the soil, leaving behind a bare bed of scarred limestone. Since then, rainwater has exploited the fissures and cracks already present in the rock, eroding away small channels, transforming the limestone pavement into a criss-crossing network of broad flat blocks, called clints, and deep vertical fissures, known as grykes.

The Burren is an enigma. It was described by one of Cromwell's officers as a 'savage land, yielding neither water enough to drown a man, nor wood enough to burn a man, nor soil enough to bury a man'. However, the grykes, fissures, shelves and clandestine pockets of the rock are abounding with plant life.

It isn't the abundance of plants that makes the Burren unique, but its floral peculiarities. Alpine, Arctic and Mediterranean plants bloom side by side throughout the national park. Maidenhair fern, usually found in warmer climates, can be found growing next to spring gentians, normally associated with high-altitude alpine meadows; the Dense-flowered Orchid, common on the Mediterranean coasts of Spain and Italy, flowers among swathes of mountain avens. The Burren is a botanical oxymoron made possible by the combination of the perfectly placed Gulf Stream and limestone's ocean-like ability to absorb heat during the summer months and then let it go over the winter. For a botanist, it is like Christmas and birthday rolled into one.

The spectacular views of the Ring of Kerry were still firmly imprinted on my mind as we set off early on the Monday for the ferry across the River Shannon. I was nervous. This was my one day in the Burren; my one day to find the Dense-flowered Orchid. Without it, I would have made a wasted trip – and, I felt, let Michael down.

I was armed with a list of sites from Sharon Parr, a plant recorder for the Botanical Society of Britain and Ireland, or BSBI for short. Welcoming and brimming with expertise, the BSBI is a charity run by volunteers who carefully record and map what's growing where and when in their local areas. The data they collect underpins the conservation of our native plants. Sharon is in charge of collating all the records for County Clare, which means her knowledge of the Burren extends down every track and gulley. If anyone knows how to find a Dense-flowered Orchid, it is Sharon.

It was an overcast, humid day. The rocky landscape perfectly matched the sky above: pewter, all-encompassing and unpredictable. We rounded a bend, narrowly avoiding a hare as it pelted across the road, to find a vast expanse of water shining pale blue. This was Lough Bunny. Clints and grykes in the limestone ran into the choppy waters of the lough. In the distance, the escarpment rose sharply into the mountainous crag of Mullagh More.

There were plants everywhere. Every crack in the limestone was sprouting green. Common bird's-foot trefoil, rue-leaved saxifrage, heath dog-violets, milkworts and hawthorn. The snowy-white flowers of mountain everlasting sprang from the pavement, spring gentians bejewelled the grass with an electric blue, and I was left speechless by the sheer number of Early Purple Orchids. There were thousands of them, speckling the slope.

Lying down on my stomach, I gazed greedily into a deep crevice and encountered a miniature jungle. Hundreds of plants thronged every crack and root-hold. There were plantains, crane's-bills, ferns, trefoils and saxifrages. Mosses and liverworts encased the smooth limestone, tiny sporophytic stalks peering upwards like periscopes. They grew over and under one another making it difficult to distinguish one plant from the next. This was chaotic, unadulterated wilderness.

'You seem to be enjoying yourself,' Michael mused, a smile spreading across his face. I was so absorbed in the abundance of species I hadn't heard him approaching. I told him that I could spend days here, weeks even. One day, eight years previously, Michael had given me a gift which had completely changed the way I observed the natural world – the camera now in my hands. For the first time he was watching me in my element.

I clambered over to a coarse lump of limestone sticking out between two bushes. Over the centuries, the rain had slowly worked away at its soluble surface, dissolving and shaping it until a hole had opened up right through the stone. This round, jagged circle in the rock provided a window for meadow vetchling, bloody crane's-bill and spring gentians. Under the nest of grasses, mosses and black lichens, a long line of ants was marching to cover: single file and military. It was like looking through a portal into a secret world.

After this excellent introduction to the Burren, we decided to move closer to Mullagh More to hunt for Dense-flowered Orchids, *Neotinea maculata*. Prominent and enduring, the softly slanting beds of Mullagh More's limestone are iconic, the Burren's picture postcard. I picked my way across the escarpment towards the foot of the hill. With limestone,

I quickly realised, you have to submit to the guidance of the grykes, the pitfall traps in the rock, and resist the impulse to wander off in arbitrary directions. The geology of the landscape determined when we turned, when we had to double back on ourselves, and whether or not we were permitted to cross onto unchartered pavements.

These hills and escarpments have hosted some of the great Irish botanists of the nineteenth century: Robert Lloyd Praeger, Ellen Hutchins, Frederick Foot and William Henry Harvey. But perhaps none so famous as Burren-born Patrick Bernard Kelly.

Kelly was a giant of a man with a thick, bushy beard and a pair of dustbin-lid hands. Fond of a pint of ale and a rich rabbit stew, Kelly was about as Irish as they come. Born in the Burren, he spent his childhood walking the hills and teaching himself botany through the observations he made in the field. By the end of the century, he had developed a name for himself as Dr P. B. O'Kelly, botanist and plant nurseryman. O'Kelly is responsible for much of what is known about the Burren's flora today, and in recognition of this his name has been immortalised: *Dactylorhiza fuchsii* var. *okellyi*, or O'Kelly's Spotted Orchid, is a white-flowered variety of Common Spotted Orchid.

O'Kelly welcomed famous botanists from across the country and guided them around the Burren. George Claridge Druce from Oxford University paid a visit; so did Robert Lloyd Praeger and Henry Levinge.

In the early 1900s, O'Kelly's nursery business grew rapidly. He published advertisements in horticulture magazines for some of the Burren's most unique wildflowers: spring gentians, grass-of-parnassus, large-flowered butterwort and pyramidal bugle. Orchids sold quickly. Two-and-six-pence for a Bee Orchid; his very own O'Kelly's Spotted Orchid went for three shillings; while the Dense-flowered Orchid – the species I was looking for – raised one-and-six-pence.

He continued dealing in plants until he was well into his eighties. When he died in 1937, he was buried near his home in Ballyvaughan, surrounded by the Burren hills where, every summer, his white orchid blooms, a wonderful memorial of his life.

Michael leaned on his walking stick and watched as I darted around, hurriedly photographing hundreds of spring plants. Many of them were new to me, all of them were utterly fascinating. Coppery, crinkle-cut fronds of rustyback fern overflowed from the shallower grykes, occasionally joined by the deep pink of bloody crane's-bill, a Burren speciality. A wider, grassier stretch of the path brought hairy rock-cress and round-leaved crane's-bill. I was rushing, though. Neither orchid I wished to see had shown up, and I was beginning to feel slightly guilty that I'd brought Michael all the way out here.

That afternoon, after a series of worrying dead-ends, we crested a hill to the sight of another lough. This one wasn't marked on my map: it was a turlough – a temporary lake that forms after periods of heavy rain, when the water table rises so high it breaches the limestone – and my final chance to see the Dense-flowered Orchid.

A rocky path, laden with spongy peat, twisted between huge glacial erratics deposited at random at the end of the last ice age. Some of these boulders were as big as Michael's car. It was humbling to imagine the forces of nature that had warped this terrain into the landscape laid before us.

I passed through a narrow gap in an old stone wall and was rewarded with a little grassy knoll that perfectly matched the description I had in my hand. Eagerly, I set about searching the short turf. The hillock was a mosaic of colour: bright-lemon tormentil, Early Purple Orchids, a cloud of mountain avens and spring gentians, sown like sapphires into the sward. But no *Neotinea*.

Frustration flooded through me.

As we continued across the Burren, stopping now and then to investigate lough shores and mossy walls, the quantity of Early Purple Orchids continued to amaze me. I had never seen so many orchids in one day, in one place, and yet no matter how hard I looked, I still couldn't find my little *Neotinea*.

I resorted to dropping in at a couple of information centres in Corofin and Kilfenora, but when I enquired about Dense-flowered Orchids, they all looked mystified and proceeded to tell me about where to see Early Purples. I was searching for too niche a plant.

In my desperation, I called Sharon Parr, willing her to pick up the phone. If she was out botanising, there would be only a slim chance of her having a signal. Six rings. Then a man answered. Unsure whether or not I had heard properly, I asked if it was Sharon. The voice on the other end of the phone said yes. Quickly covering my surprise, I adopted the manner of someone who knew that men could be called Sharon and briefly explained my predicament. He, for I was convinced Sharon was a 'he', directed me to various places on the north-west coast where there had been recent reports of *Neotinea* in flower.

I ended the call, grateful for the information but somewhat bemused. Perhaps Sharon was a unisex name in Ireland. It seemed like it could be a very Irish thing to do, call your son Sharon. So I accepted it and we drove on. Michael's patience and understanding that day were immense. A relaxing venture into the Burren to see a couple of new orchids had rapidly turned into a wild-goose chase across the whole of Clare, a drive of heroic standards for an arthritic seventy-year-old.

We eventually arrived at the coast, but the sites Sharon had dictated hurriedly over the phone proved difficult to find. At the lighthouse near Ballyvaughan, Patrick O'Kelly's old hunting ground, I searched meticulously among the small boulders and smoothed patches of limestone. More spring gentians and some bright Irish saxifrage, but not what I was looking for. I was bitter and disheartened. I'd found everything I'd hoped for except the two orchids I had specifically come to see.

Four hours later, we sat watching the sun set softly over the river back in Ballybunion. I checked my email and found a reply from Steve: the orchids I had sent him were a variety of Irish Marsh called *kerryensis* which was known to grow on the south-west coast of Ireland and have paler flowers and unspotted leaves. An unusual variety, but they were still Irish Marsh Orchids; they still counted. The Dense-flowered Orchid, however, I had failed to find, which meant I would have to come back to the Burren to avoid failing in my venture. Another flight, and more time away from orchids in England.

Stumped by Ireland's Mediterranean Orchid

I spent the last couple of days with Michael recovering from our adventure in the Burren. We visited the pretty village of Adare, with its quaint streets, medieval monasteries and thatched cottages; we got stuck for an hour in Listowel while waiting for stragglers in an awfully managed cycling event, leaving Michael wondering 'what the hell goes through Irish heads'; and we spent a lot of time lounging around in front of the TV in the cottage, laughing at the contestants competing in the Eurovision song contest.

On my final evening, we went down to the cliffs near Ballybunion and watched fulmars swoop and circle, seemingly just for fun, and choughs that pierced the quiet with sharp cries as they arrowed past. A light breeze carried the sound of waves crashing against the rocks far below as we ambled slowly along the cliff.

As we walked, Michael broached a topic we had so far avoided: my love life. 'So do your girlfriends know about this orchid thing?' he pried, eyes twinkling mischievously. I sighed. I was about as near to having a girlfriend as we'd been to finding Dense-flowered Orchids. Attractive, orchid-inclined girls are hard to come by.

Throughout my time at school I'd been constantly worried that the girls I fancied would find out about my botanising. I felt certain I would be ridiculed for it and as a result would never have a chance with any of them. If any of them found out it would be torturously embarrassing.

Spying a promising area of reeds in the adjacent field, I took advantage of a broken wire fence to escape the uncomfortable conversation. I scurried over, hoping to spring another, more convincing Irish Marsh Orchid. Pond water-crowfoot ran rampantly across the dry, hoof-trampled ground. As I got down on my knees to look more closely, a flash of pink caught the corner of my eye, but it was only another *kerryensis* marsh orchid. I desperately wanted to see a typical one.

On the verge of giving up, reassuring myself that I would probably see one if I returned to the Burren, I walked the last ditch back to the car. Halfway down, I stopped and smiled. There, in a warm, sheltered hollow, was the glorious deep pink of a fully flowering Irish Marsh Orchid. Its petals were smooth and velvety; the colours unphotographably vibrant. Equally satisfying was the ease with which I had been able to identify it, pure and unambiguous.

I returned to the car where Michael sat waiting for me. He took one look and a smile spread across his face. 'What have you seen?' he said. I explained, sharing the joy I'd felt at finding that perfect Irish Marsh Orchid. The disappointment of the previous day seemed a long way away.

Michael drove me to the airport the following morning. I remember talking to him about the book I wanted to write about my orchid hunt; I asked him if, when the time came, he would be willing to proofread some of my chapters. His face glowed with delight.

Upon arriving at Shannon airport, he waved me off, leaning on his walking stick and standing out from the crowd in his bright-purple jumper. I didn't know it then, but that was the last time I ever saw Michael. He passed away the following year, and I cannot express how grateful I am for that week in Ireland with him. This chapter is a dedication to him, and the time we spent together.

3

The Loose Flowers of Jersey

*'Historically, the search for and acquisition of orchids has
been one of the most manly of feats.'*

Richard Leighton, *Orchid Hunting
in the Florida Everglades* (2012)

Jersey
May 2013

Men who have a passion for plants have often been typecast
as effeminate or gay. Why, though, is an interest in flowers
deemed feminine? Surely botany is just another of the natural
sciences. An interest in mountains, or indeed the bears that
live on them, is hardly considered emasculating. I think we all
have an innate connection to nature; it's just that many of us
aren't aware of it and don't tap into it. An appreciation of our
wild world is a fundamental part of what it is to be human,
regardless of one's gender or sexual preference.

At the start of my next expedition, I had a long and

unexpected conversation on these matters with a taxi driver in Jersey which got me thinking. Did people assume I was gay when I sheepishly outlined the challenge I'd set myself? As a nineteen-year-old who hadn't been in a proper relationship before, I wasn't exactly dispelling the notion. I didn't want girls to mistake me for something I'm not, otherwise I'd never stand a chance.

I had come to the Channel Islands to find Loose-flowered Orchids: tall, pink, supermodel-like plants more often encountered in the Mediterranean than in Britain. As this orchid is no longer found growing naturally on the mainland, Jersey and Guernsey were my only options for seeing truly wild examples of it.

I'd touched down at Gatwick the night before, exhausted and slightly defeated by my time in Ireland. It was only as I walked over to baggage reclaim that I realised I'd forgotten to book a hotel at the airport. My connection to Jersey the following morning was too early for me to go home. I tried to find a hotel with a room, to no avail, and after one sad phone call with my mother, I'd resigned myself to the fact that I would have to sit out the night in the airport.

Settling down on a seat overlooking the arrivals lounge with a coffee, a bag of foam banana sweets and my iPod, I began people watching. It seemed the world was arriving in London – thousands of people, Belizean, Japanese, Ethiopian, Latvian and Canadian. At around four, two armed policemen with the biggest guns I'd ever seen sauntered through our slumbering group. I tried my best to look nonchalant.

The night stretched on, until eventually the clock struck six and I made my way over to the departure gate. The next few hours passed in a blur. After our conversation, the taxi driver dropped me off at my B&B. I staggered up the steps with my luggage and headed to reception where Kerry, the owner, welcomed me as Lewis Borstadan, and wished me a pleasant time in Jersey.

Over the next few days, I explored Jersey, enjoying the beach and the top-floor suite I'd been upgraded to, giving the orchids a few more days to reach their peak. The Channel Islands are actually situated much closer to the coast of France than to Britain. Technically British Crown dependencies rather than part of Britain per se, they're made up of seven permanently inhabited islands: Jersey, Guernsey, Alderney, Sark, Herm, Jethou and Brecqhou. Puzzle-piece shaped, Jersey is the largest of all the islands and is renowned for its sweeping clifftop views, cultural history and picturesque seaside towns.

Each morning, I piled heaven-sent scrambled egg onto my plate in preparation for hours of exploration. Once, Kerry walked past while I had an extremely large mouthful of egg and bacon and trilled: 'Morning, Lewis, I hope you slept well!' With a name like Leif, I'd had a lifetime of being called all sorts of things: Leaf, Life, Layif. My parents have had conversations with people who are unable to believe the coincidence that their son Leaf is so interested in plants. I was even called Loaf once. Lewis, however, was not on the list.

I started to look around the island: I visited Elizabeth Castle, a sixteenth-century fortress cut off from the capital St Helier by the tide twice a day; I took long walks along panoramic beaches at St Brelade's Bay and St Aubin; and explored the rocky coastal paths leading to sheltered, secluded coves.

I hired a bike from the B&B and cycled further afield, battling the steep roads up from the coast without any gears. I passed fields of tan-brown Jersey cattle, stopping to stroke their friendly, inquisitive muzzles. Before the turn of the nineteenth century, cows would be offered as dowry for marriages between Guernsey and Jersey islanders.

When the time came to go in search of Loose-flowered Orchids, I had relaxed into life on Jersey and had familiarised myself with the coastline. I took the bike out and pedalled up a windy lane to the top of the cliffs. Tiny purple-and-yellow snapdragon flowers of ivy-leaved toadflax tumbled from crevices in the old walls.

In the Channel Islands, the Loose-flowered Orchid is a plant of boggy pastures and water meadows. It was first recorded on Jersey by C. C. Babington in 1837. While also present on Guernsey, it is inexplicably absent from the other islands in the group. There are plenty of historical accounts of its wide distribution across both Jersey and Guernsey. E.D. Marquand, upon visiting north-west Guernsey in 1901, reported that 'at the beginning of June the fields are quite purple with these beautiful flowers'; other plant hunters in the same period reported that at times it was almost impossible to walk without trampling them. Drainage, development and a shift in agricultural methods have been catastrophic for populations, with many just a fraction of their former selves. If it hadn't been for a significant conservation effort from the Jersey National Trust, we may have lost this species altogether.

Jersey is a maze of so-called Green Lanes – roads prioritised for pedestrians and cyclists. And it seemed to me, as I pedalled along, that without the smog of exhaust fumes and perpetual traffic jams, nature had taken control. Every nook and cranny in the crumbling stone walls was overflowing with small polypody ferns, herb-robert, stonecrops and ghostly green navelwort. Some were awash with Mexican fleabane, a dainty garden escapee, while others were full of lichens: crimson goblets and ragged wizard-beards. Orange-tip butterflies were zigzagging up and down the roads, passing from garlic mustard to three-cornered garlic and back again.

I sped down the hill towards the marshland behind the dunes in Les Mielles Nature Reserve, a yellow blur of gorse whizzing past on both sides. The lanes periodically passed through small hamlets with stalls and honesty boxes offering freshly dug Jersey Royal potatoes. I was ticking off stereotypes faster than orchids.

Passenger planes drifted overhead into the airport as I skidded to a halt outside the entrance to Le Noir Pré, known locally as the Orchid Field. Purchased by the Jersey National Trust in 1972 on the recommendation of Frances Le Sueur, a distinguished local botanist, Le Noir Pré had been one of many meadows under threat of ploughing for potato crops.

Le Noir Pré is actually made up of two fields, both damp and with a sea of reeds in their centre. That day they were speckled yellow with buttercups, the white umbels of hemlock water-dropwort lined the edges, almost masking the sign next to the gate that read 'Keep out – reserve open on the 26th May onwards'. It was 25th May. But I hadn't come all this way to be kept out by a sign. After a quick glance around, I scaled the gate and jumped down on the other side.

I spent the next hour and a half in among a huge colony of Loose-flowered Orchids. All around me rose lofty spikes with rich, velvety purple flowers, each with a sharply folded lip and rabbit-ear sepals held aloft. They looked like stretched-out Early Purple Orchids, but more spectacular. Le Noir Pré was a catwalk for these supermodel orchids. Dressed in a royal magenta, they practically strutted. The Latin species name, *laxiflora*, provides a fitting description of the way this sparsely arranged inflorescence rises tall and stately from the surrounding vegetation.

While Loose-flowered is an apt description of this plant, it is also known by various other names: locally it is the Jersey Orchid, and in France 'des pentecôtes', a reference to its time of flowering. I texted my parents to tell them I'd found the Pentecost Orchid and my father replied: 'It's a week late!'

I hadn't expected to find such a profusion. According to Babington, this was a common scene in the nineteenth century, as it was found in 'almost every wet meadow and bog' in Jersey and Guernsey. Aside from a small but thriving population in Sussex, introduced using seeds collected in Crete, there are sparse records of *laxiflora* in mainland Britain. Weirdly, most are from County Durham in the late 1800s.

Five minutes of walking around the meadow and I had spotted the slightly softer pink of three Southern Marsh Orchids. Each spike was densely packed with buds that were mostly still planning and scheming. Beneath each bud protruded a thin, pointed leaf-like structure called a bract, making the plant look like a green mace. The saucer-like lips of the few flowers that had opened were still crinkled, like the rumpled wings of a freshly emerged butterfly. They were disconcertingly early. Normally I would expect them to blossom in the middle of June. But I was in Jersey, I reminded myself, considerably further south than the Wiltshire countryside I was accustomed to.

Overlooked until the early twentieth century, the South-ern Marsh Orchid was named *Dactylorhiza praetermissa* by G. C. Druce in 1914. Since then it has been widely recognised as the commonest marsh orchid in the south of England. In his landmark work *Wild Orchids of Britain* in 1951, V. S. Sum-merhayes, curator of the orchid herbarium at Kew Gardens for thirty-nine years, notes that it can be found growing in a 'variety of wet places, damp meadows, water meadows, lowland peat-bogs, fens, and marshes among sand dunes'.

As the breeze dropped and the heady fragrance of orchids filled the air, I paused, struck by the mood fashioned by the intense colour of the colony: not spring, but decadent mid-summer; the romantic, hopeful flush of pleasure associated with pink. Loose-flowered and Southern Marsh Orchids blended into the blush of red campion, dog-rose and rag-ged-robin, as well as crane's-bills and stork's-bills, vetches and pimpernels, all displaying their own unique shades of pinks and reds.

After Le Noir Pré, I spent a couple of hours exploring the fields surrounding St Ouen's Pond. I found more Loose-flow-ered Orchids there, their deep-purple flowers contrasting with the pure white of common star-of-bethlehem. As I crossed to the drier ground, I found spring sandwort and, much to my delight, the fluffy lilac bobs of hare's-tail grass. Overhead a marsh harrier was keening, circling the pond and surrounding reed beds. I followed its shadow as it passed, a black silhouette against a bright floral background.

A bit further on and there was my third new species of the day: Common Spotted Orchids. Like the marsh orchids, they were very early – I was used to seeing these hit their peak around the summer solstice back home – and there were droves of them. Their waxy leaves are splattered with purple, and their flowers are pale fuchsia, with lilac loops and cryptic squiggles, as if telling a story. I kept my distance: get too close and they're hypnotic.

I jumped back on my bike and cycled towards the sea, passing a long ditch of Loose-flowered Orchids. They were certainly the most southerly orchids I would find this year. The Channel Islands have played host to several other orchids in the past that haven't yet established themselves on the British mainland. Tongue Orchids, of the genus *Serapias*, are Mediterranean in origin but occasionally turn up on the Channel Islands. They are pale grey with reddish streaks and a flat auburn tongue of a petal. They supposedly attract pollinators by mimicking little insect burrows.

The name *Serapias* is derived from *Serapis*, a composite Graeco-Egyptian god uniting the sacred bull *Apis* with *Osiris*, ruler of the underworld. His orgiastic cult inspired the Greek physician Pedanius Dioscorides to give the name *Serapias* to an aphrodisiacal orchid, which was later applied to the genus by Linnaeus in 1753.

Until the late 1980s, there were no records at all of these species in the UK, but in 1989 two spikes of Small-flowered Tongue Orchid were found on the coast of Cornwall. Since then, numbers have oscillated between zero and five each year. It's unknown whether they were established as a result of wind-blown seed or deliberately introduced by a local orchid fanatic. But the fact they survived and flowered suggests global warming may already be having an effect on the range of orchid species. Might we see an influx of new species over the course of the next century? Will sightings of Small-flowered Tongue Orchid become more frequent?

I arrived at the beach, locked my bike to a railing and climbed a dune through swathes of thrift and sea campion. At the top, I had a wide-angled, unadulterated view of a crisp, docile sea, its shallow waves reflecting the sun in rhythmic pulses. Down on the shoreline, three wading birds were picking their way along the beach, acting in tandem with the water and rushing to safer ground with every swell and surge. A light breeze caused tiny grains of sand to swarm across the path before settling among the marram at my feet. The smell of salt was left hanging in the air.

After a long, refreshing swim in the sea, I returned to the dunes. I had really been in my element in the fields with the Loose-flowered Orchids. This was an ungendered obsession: with the natural world, with orchids, with nature in its purest form.

4

The Lady of the Woods and Her Hanged Man

'And it must be a never-forgotten delight for anybody but the most arid of Peter Bells to come upon the Lady Orchid for the first time, its exotic spikes towering regally above the bowed, respectful bluebells in some Kentish copse.'

Jocelyn Brooke, *The Wild Orchids of Britain* (1950)

Kent
May 2013

Jocelyn Brooke considered British orchid enthusiasts to be few but fanatical. 'Their hobby is a cult, a kind of freemasonry with all the jealously guarded secrets of such institutions,' he wrote in *The Wild Orchids of Britain*, one of the landmark volumes on our native orchids. In it he advises that 'if one were asked to introduce the untutored amateur to the native orchids, probably one's wisest course would be to make for one of the less-frequented downland districts in Kent, say, or on the Hampshire and Sussex border'. It is here, he muses, on open downland or in the shade of the woods, that the orchid hunter is most likely to come across large colonies of orchids.

Kent is renowned for being the orchid capital of the UK, hosting well over half of our native species, including some that are found nowhere else in the country. Plant distributions are affected by a whole host of different factors. Is the ground wet or dry? Can it tolerate disturbance? How well does it deal with shade, grazing or the presence of other species? Orchids can be highly sensitive to light. The Bird's-nest Orchid and Violet Helleborine grow in the darkest of woods, where the sun-loving orchids of the woodland rides are never found.

The varied geology of the British Isles and the distribution of our orchids are fundamentally linked, sometimes so closely that knowing the underlying rock type can help pin down an identification. Much of Kent is composed of alkaline soil derived from the underlying chalk. The orchids found in Kent love calcium. Contrastingly, the presence of calcium in the soil makes it difficult for other species, for example those found in the bogs and wet pastures of the north and west of the country, to take up essential nutrients. Orchids are extremely fussy; each species has its own set of specific requirements that enables it to grow and reproduce.

The calcium-loving orchids of Kent prefer old woodland or undisturbed, well-grazed downland. The county has an abundance of rare species. Even today it is not unusual to walk out onto the North Downs and come across large colonies of orchids, although how long it will remain this way is yet to be seen. In the last sixty years, grazing practices in Britain have been in decline; as a result, robust, nutrient-hogging grasses have been allowed to dominate, and are outcompeting the more delicate orchids. Because animals are no longer grazing the pastures, scrub has been allowed to grow in many places and once open downlands have become miniature forests of hawthorn and ash saplings.

The rich downland of Kent, on the other hand, which remains pocketed away in chalky valleys, provides little havens for a multitude of orchid species. It was here, on a sunny Monday afternoon in May, that I undertook the next stage of my journey. I spent the morning packing the car and fending off the hitch-hiking efforts of our cat, Tabitha, who seemed determined to come with me. Extrication from the car's footwell or my open bag only made her more resolute. At last I was underway, heading on mercifully traffic-free roads across the south of England to the North Downs. I let the windows down and sang 'Hooked on a Feeling' at the top of my lungs. A Waitrose delivery lorry flashed past and I grinned smugly as I thought of my work colleagues stuck indoors stacking shelves of juice.

I pulled up outside Darland Banks, a Kent Wildlife Trust reserve tucked away between Gillingham and the M2. This was where I was hoping to find the Man Orchid, *Orchis anthropophora*. I'd never been able to search for this one before, as its flowering time unfortunately coincided with school exam periods.

Darland Banks is a jewel in the crown of the Kentish countryside. Out on the escarpments, fresh grass was being blown into swirls by the breeze, and the warm sun was shining down on the lemon-yellow fields of oil seed rape. Narrow chalky paths wound their way across the hillside, then disappeared enticingly into the distance.

I moseyed slowly along the bank, exchanging nods of acknowledgement with dog walkers and fellow nature enthusiasts. Some were clearly here for the butterflies, looking for skippers or the flash of a green hairstreak. Others, and it was harder to confidently pick them out, were almost certainly orchid hunting. A woman shuffled towards me, her eyes glued to the ground. At the top of the hill, a couple of old blokes wearing khaki jackets and wide-brimmed hats were scanning the grassland with methodical precision. These were the botanists, the orchidophiles, in their natural habitat. I joined in, mirroring their purposeful gait; eyes sweeping between milkworts, field madder and the golden rings of horseshoe vetch.

The Man Orchid has been given a variety of different scientific names. Linnaeus originally placed it in the insect-mimicking genus *Ophrys*. Here it remained until it was unceremoniously moved to *Aceras* and then eventually to *Orchis*, where it joined other similarly anthropomorphic species, whose flowers resemble little human figures. Classification can be a real head-scratcher. Over the years, it has been a source of division, pitting new-fangled molecular biology against years of taxonomic tradition.

The sun had dipped behind a small cloud, illuminating its wispy exterior and sending light cascading in all directions. With no luck on the upper slopes, I dawdled down to the bottom of the hill into an area recently cleared of scrub. Tiny hawthorns pricked through the tufts of grass. According to Summerhayes, '*Aceras* differs from many other orchids in being often found mainly or exclusively near the base of slopes, though this is not always the case'. He also notes that 'on the whole, the Man Orchid grows best in open situations, and does not appear tolerant of much shade. Sometimes plants may be found growing in the shelter of isolated bushes or among very open scrub'. This, then, seemed like a spot they might like.

I had been walking for fifteen minutes before I came across my first Man Orchid. Even after a decade of plant hunting, the thrill of a find never wanes. And on this occasion I was pleasantly surprised by how quickly I had managed to spot this subtle, limey-yellow orchid. The slender spike was crowded with an asparagus tip of flowers, each one shaped into a remarkable little man – not unlike the figure on traffic light crossings. Green sepals and petals formed the head and the yellow lip hanging down looked just like the body. The arms and legs were a beautiful burgundy.

Each miniature figurine was different. Some were standing stock still, held to eternal attention. Others were frozen in a perpetual stride, while a disturbing number hung limply as if on the gallows. Colloquial English names for this plant are rare, but in France it is aptly, if morbidly, named *L'Homme Pendu*, 'the hanged man'. In Germany it is referred to as *Puppenorchis* and in Italy as *Ballerino*. This last one I thought a bit far-fetched until I came across one spike of nine flowers, all choreographed into a mesmerising yellow ballet.

The first British record for the Man Orchid was made in the late 1600s, in a disused gravel pit in Essex by a Mr Dale. Since then it has been discovered across the south-east, with the majority of populations in Kent and Surrey. It is less frequent to the north and west. Classified as Endangered, it is a localised species and unlikely to be stumbled upon outside of nature reserves.

I moved to another part of the bank, about halfway up the hill where wild columbine rose, dogwood and the white blooms of the wayfaring-tree grew among the hedges. The plants here were more typical of those you'd find growing alongside chalk-loving orchids, with swathes of downy oat-grass, kidney vetch and yellow rattle. As far as I could tell, though, no more orchids.

Summerhayes notes that the Man Orchid is a somewhat inconspicuous species which might easily remain unnoticed in an area for a relatively long time. I decided to put this to the test, sitting myself down in the hollow of an old rabbit hole. I waited. At first, nothing happened; the palette of colours remained the same, different shades intermingling with one another as the plants shifted in the breeze. But after a few minutes, I began to notice Man Orchids; they were lurking behind vetches and buttercups, peering at me from their hiding places. Once I had trained my eye to see them, they appeared more and more frequently, and reached further and further from where I sat, like shy children coming out to inspect their parents' dinner guests.

I had been sitting just off the path but jumped when I felt the inquisitive snuffling of a Labrador's nose on the back of my neck. So absorbed in orchid hunting, I hadn't heard its approach, nor its owners, who were only a few metres behind. After a short and somewhat stilted conversation, during which the Labrador attempted to sniff at my crotch, they told me that most of the Men were at the other end of the reserve. Local knowledge was exactly what I needed.

I thanked them for their time and set about gathering my things. Unfortunately, we were now headed in the same direction and so I awkwardly followed them as they made their way along the hillside, bending down to examine a plant whenever they chose to look back. We continued in this way for some time, stopping and starting along the mazy pathways. As we walked, I began seeing Man Orchids in the grass with increasing frequency. Up until then, I had seen perhaps twenty plants and assumed I was too early for so rare a flower, but I was now closing in on fifty.

Two kissing gates later and there were thousands of them carpeting the slopes. I stopped and counted thirty-three plants in less than a square metre. Every time I thought I had found the perfect plant to photograph a better one appeared over the top of my lens.

I wandered slowly and carefully through the colony, picking my way with utmost respect. Up ahead was a spike so big and pale that I initially took it to be wild mignonette, but as I drew closer its true identity became clear and my eyes widened. This Goliath of a specimen was forty centimetres tall and boasted more than fifty flowers, another fifty ready to spring from their buds at the first sign of danger. It smelled, bizarrely, of frying onions.

These orchids were paler in colour, each Man a familiar golden yellow but without the burgundy limbs of the plants I had seen earlier. I was once again hooked, finding that I had to study each and every flower: all unique, all alive. Two here were angled towards one another, caught in a never-ending conversation, and on closer inspection I noticed that some had a little protuberance between their legs: a confirmation of their manhood. J. E. Lousley, writing about the Man Orchid in *Wild Flowers of Chalk & Limestone* (1950), postulated that there is 'an alternative resemblance to the human race which amuses people with imagination. The hood may be regarded as portraying the head of a dwarf with the lip for his flowing beard'. I could see what he meant, though could not rid my mind of its resemblance to a hanged man.

Hot but giddy with success, I reluctantly made my way back to the car. The melodious echo of a thrush floated down from the trees, momentarily drowning out the bland drone of the motorway in the distance. The late-afternoon sun had turned the hillside a golden green, much to the delight of the butterflies which were still bumbling from flower to flower. A breeze swept across the slope, making the ox-eye daisies bob about in mutual agreement. I wanted more than anything to spend the evening at Darland Banks and watch the sun dip below the horizon as the day came to a close, but I was being called further into Kent where, rumour had it, the Lady of the Woods was waiting.

I was still marvelling at the intricacy of the Man Orchids as I pulled off the motorway and began following the narrow lanes that wound their way through the Kentish countryside to Denge Wood. Heavy rain had been forecast for the next day and I knew that realistically this was my only chance to see Lady Orchids. Despite their names, these orchids are different species, rather than different sexes.

The Lady of the Woods, or Maid of Kent as she is locally known, was first discovered in Britain in 1666. Like so many others, she has suffered at the hands of collectors. Preyed upon for her beauty, the Lady Orchid has been forced to hide deep in the Kentish countryside, taking refuge in small copses and ancient beech woods. It's here she remains to this day, protected by a select few dedicated guardians and attracting hundreds of admirers every year. One of her faithful guardians was Jocelyn Brooke, who described the Lady of the Woods as 'orchid royalty' for her rarity and beauty, placing her on a throne alongside the Military and Monkey Orchids.

Born in 1908, Brooke grew up in Kent and quickly developed precocious interests in botany and fireworks. Each week he would use his pocket money to purchase little packets of strontium nitrate, potassium chlorate from the local chemist. After fruitless days hunting for rare orchids he consoled himself by building firecrackers and rainbow rockets. In his acclaimed book *The Military Orchid*, he recalls how 'my mother lived in hourly terror of my blowing myself and the entire family to smithereens. Once or twice I nearly did'.

As far as botany was concerned, what started as a well-rounded curiosity soon became a specialist interest as he zeroed in on the Orchidaceae. He was attracted by 'those floral aristocrats, with their equivocal air of belonging partly to the vegetable [and] partly to the animal kingdom'.

By the time he was seven, he was already a keen botanist and had been 'bitten with orchid-mania'. For his birthday he received *British Orchids, How to Tell One from Another* by Colonel J. S. F. Mackenzie to supplement his other wildflower guides. While admitting that the colonel's book was not the best introduction to its subject, Brooke assures us that he was a 'true orchidomane' and clearly held him in high esteem.

In his writing, Brooke mentions a Mr Bundock, whose job it was to empty the composting toilet at the cottage where he and his family would spend the summer. Mr Bundock had previously provided very little interest to the young Brooke, but this all changed with a passing remark about the discovery of a local Lizard Orchid. Mr Bundock promised to bring the boy back some specimens the next evening, and Brooke waited in anticipation, only to be severely disappointed when Mr Bundock presented him with several Man Orchids, a species he had already found himself. But his disappointment, he writes, was 'mitigated by the other orchid which Mr Bundock had brought me. This was unfamiliar: a tall, handsome spike of purple-brown and pink-spotted flowers. Obviously, I thought, it came under the desirable category of Very Rare Orchids. But which was it?'

Brooke was determined to identify Mr Bundock's new orchid. Equipped with his new books, he set about tackling this problem, only to find that the authors contradicted one another. He desperately wanted his specimen to be a Military Orchid (*And yet... and yet... if only it could be* Orchis militaris.'), and it was identified as such by Colonel Mackenzie – but according to Edward Step, the plant was far more likely to be a Lady Orchid, colloquially known as the Great Brown-winged Orchid, which had failed to receive even a mention in the colonel's book. Torn between the two, he wanted to believe that it was a Military, but in the end he reluctantly admitted that Mr Bundock's orchid was the Great Brown-winged – 'Edward Step, after all, could hardly have invented *Orchis purpurea* [Lady Orchid] out of sheer malice'. For now, the Military Orchid remained unfound.

Denge Wood is a large woodland complex south-west of Canterbury. Owned by the Forestry Commission and the Woodland Trust, it remains one of the best sites in Kent for Lady Orchids. But you could easily spend days walking around the wood in May and never come across this floral aristocrat, so concealed are its colonies. The 'x' on my map seemed to shrink into insignificance as I tramped down a dusty dirt track. Above me a dome of beech leaves dappled and sparkled. Each leaf was gently unfolding like the crinkled wings of a newly hatched dragonfly.

I was practically running by the time I arrived at the site, anxious to find the orchids before the evening light faded. In my haste, I almost missed the gate by the side of the track. Rushing through, I stumbled down a short slope and into a clearing. There, where beech and ash gave way to wayfaring-tree and stunted pines, the Lady Orchids had made their home.

Beneath a bonnet of maroon, the white lip of each flower forms a tiny figure dressed in a billowing blouse with purple tufts of miniscule hairs. Lousley describes her as having 'an outline recalling the sketches which Victorian children used to draw of crinolined ladies, and the deep brown hood may be likened to their bonnets'. I personally think the adjective 'brown' is used far too often for a colour that I would liken to claret or, as the Latin epithet *purpurea* suggests, a rich reddish purple. She does have a claim to royalty after all.

Kent is the rightful home of Britain's Lady Orchids – as Lousley puts it, the Lady is 'a specially Kentish plant' – and although rare, where found they often occur in large numbers. Here in Denge Wood I had several hundred plants all to myself, a pantheon of purple and white. But in recent years, this orchid has plucked up the courage to begin exploring westward. She has been spotted in the woods on Salisbury Plain and admiring the view over the Thames in Oxfordshire. According to early records, the Lady may not have been so rare outside Kent. Anne Pratt, an immensely energetic nineteenth century botanist, observes that the orchid was 'often carried into the towns in baskets for sale, mingling among green Tway-blades, and dim brown Bird's-nests, and overhung by graceful ferns'. Nearly one hundred years later, Jocelyn Brooke brands this 'a regrettable practice which still, unfortunately, survives among the local hawkers, who sell this beautiful orchid at the street corners (at sixpence a bunch)'.

Anne Pratt (1806–1893) was not just a fantastic botanist, but also one of the best English botanical illustrators of the Victorian era. She is variously described as being sensible and humorous, displaying a complete lack of fear for gamekeepers, barbed-wire fences and the weather.

It must have been a lot of fun plant hunting with Miss Pratt. Jocelyn Brooke writes about her tireless botanising in *The Military Orchid*: 'One imagines her setting forth on some summer's afternoon in the 1850s, perhaps escorted by some frock-coated clergyman... sensibly clad, minutely observant, humorously deprecating the vestiges of superstition among the villagers, and always ready, by an appropriate word here or there, to assist in spreading the Light of Revelation. Toiling over the Dover Cliffs for *Silene nutans* [Nottingham catch-fly], wading through the marshes about Sandwich for the greater spearwort, or searching "in the woodlands or on the bushy hill" for *Orchis purpurea* [Lady Orchid] – one sees her, indomitable but incurably lady-like, pursuing her purposeful way through the Kentish countryside, her tweeded figure bathed in the warm, golden light of a Victorian Sunday afternoon in summer'.

I drifted through the clearing, from one council of lofty ladies to another. It's one of the largest orchids to grow in Britain. I then began to notice the subtle differences between each plant. Some had pale bubblegum pinks and rose-tinted veils, while others were completely white and freckle-less, hooded by pastel-green sepals. A significant proportion of the colony towered elegantly above the surrounding herbs, their flowers showy and flamboyant, but a few had stunted, underdeveloped stems.

I couldn't help but compare their florets to the Man Orchids I had seen earlier in the afternoon, and found myself equally enthralled by these tiny white figurines. They were less active than the Men, preferring to stand tall and be admired by their onlookers than dance around like the *Ballerinos*.

Lady Orchids aren't always as dignified as they may seem though. At the turn of the millenium a Lady Orchid was discovered at Hartslock Nature Reserve in Oxfordshire, growing among a small population of her close relative the Monkey Orchid. It turns out they couldn't keep their hands off each other. Each spring, on the small hillside overlooking the Thames, the Lady and the Monkey undergo a coupling every bit as incestuous as that practised by the royal families of ancient Egypt. Lady Orchids hybridise at any opportunity, running riot across western Europe with Military, Monkey and Man Orchids, forming extensive broods of illegitimate offspring.

Denge Wood was slowly succumbing to the shadows. I was still taking photos, although it was becoming increasingly difficult in the low light. Making my way past one last congregation of Ladies, I spotted the dull green of two Common Twayblades, my tenth orchid of the year, nestled among the dog's-mercury. Their slender spikes bear many inconspicuous green flowers, giving them the remarkable ability to disappear into the background while in plain sight.

The Twayblade has been known to grow in Britain for

centuries, the earliest record being published in 1548 in Turner's *Names of Herbes*. He calls it the 'Martagon', says it is found 'in many places of Englande in watery middowes and in woddes'. It is second only to the Common Spotted as Britain's most abundant orchid, although I suspect its extremely camouflaged flowers are considerably under-recorded, particularly given its affection for hiding in 'woddes'.

Few local names for this species are recorded, but in Wiltshire it is colloquially known as Adder's Tongue – the lip is forked and not unlike a reptilian tongue – while in Somerset the name Sweethearts is often used, because of its pair of distinctive egg-shaped leaves. Gerard describes each flower as 'resembling a gnat, or a little gosling newly hatched', but no matter what angle I looked at it, I couldn't understand where he was coming from.

Once again, Brooke has plenty to say about this species: 'If the orchids represent the "Royal Family" of flowering plants, what are we to say of the undistinguished and far-from-regal Twayblade? Its royalty seems nominal and arbitrary – a commoner ennobled, so to speak... scarcely, indeed, ennobled at all... Yet the humble, plebeian Twayblade is a true orchid, and, though outwardly unimpressive, has more than its fair share of the structural oddity of the family. Darwin, in fact, declared it to be "one of the most remarkable of the whole order". It almost seems as though this outward unattractiveness were compensated by the ingenuity of the plant's sexual mechanisms.'

Pollination is indeed remarkable in the Twayblade, illustrating how beautifully balanced the relationship between plants and insects can be. The lip, hanging down below the other sepals and petals, has a narrow nectar-producing groove, which the insect follows until it comes into contact with a tongue-like structure called the rostellum. Contact with the rostellum triggers the release of a drop of sticky liquid which glues the pollinia onto the insect's head in seconds. The insect is startled by the sudden movement and flies away, often to another plant, where it once again begins feeding on nectar, simultaneously bringing the pollinia into contact with the sticky stigma. Never underestimate a Twayblade.

By now the sun had disappeared completely, leaving behind a pink blur above the trees. The stately Ladies of Denge Wood stood tall all around me, watched over by attentive Twayblades and serenaded by the harmonious song of the blackbird and the song thrush. Jocelyn Brooke was no stranger to the serenity of Kentish woodland and no doubt he spent countless hours searching for precious spikes of *purpurea* and *militaris*. A year or two later, still searching for the elusive Military orchid, he was once again presented with a fresh specimen. With more detailed information about the whereabouts of this flower, he eventually managed to find it himself and confirms, having long ago abandoned the idea that *militaris* walks in the woods, that it was the Great Brown-winged. He goes on to provide a wonderful description of it as 'the most regal of British orchids, and perhaps the loveliest of English wildflowers: its tall pagodas of brown-hooded, white-lipped blossoms towering grandly, like some alien visitor, exotic as Miss Trumpett [a sophisticated local lady] at a village tea party, above the fading bluebells and the drab thickets of dog's mercury, in a wood which I had known all my childhood, but whose distinguished inhabitant I had never before discovered'.

5

Orchids of the Emerald Isle

*'For if delight may provoke mens labor, what greater
delight is there than to behold the earth apparelled with
plants, as with a robe of embroidered worke, set with
Orient pearles, and garnished with great diversitie of rare
and costly jewels?'*

John Gerard, *The Herball* or *General
Historie of Plantes* (1597)

Ireland
May 2013

The next day dawned wet. I'd spent the night at a youth hostel
in Canterbury amid a coach load of Dutch school children
and had woken early, eager to get out before the onslaught
on the canteen.

I spent the morning exploring the cobbled streets of Canterbury, ducking into small shops and independent cafés that
on a sunny day would be bustling with people. Outside the

cathedral, I watched as bedraggled American tourists handed over their fresh bank notes, their 'I heart London' T-shirts hidden beneath coats and umbrellas. The weather was grim, and I was relieved that I had made the effort to squeeze in the Lady Orchids the evening before, as photographing them in the rain would have been cold and dissatisfying.

I'd decided to gamble on a return to Ireland to have another go at finding the Dense-flowered Orchid. Thanks to the late spring, it would still be flowering and I was determined not to fail so early in the season. To increase my chances of success I enlisted the help of Sharon Parr, the BSBI county recorder for County Clare, who had provided me with directions on my previous visit. During our conversation on the phone, it became clear that there was no guarantee we'd find one, but it was worth risking another return flight.

So that evening, I flew from Gatwick to Shannon, for the second time in as many weeks. I paid a jolly Irish taxi driver an extortionate amount of money for a short drive from the airport to the small town of Sixmilebridge. I had booked a B&B for the next two nights based on the optimistic assumption that I'd only need a day to make my find.

As the taxi pulled away into the darkness, I turned to the brash red door, its paint peeling like bad sunburn. There were two doorbells mounted on the wall so I pressed the first, listening to the dull chime as it echoed off the walls inside. Nothing. I tried the other and stepped back with shock as a static shriek erupted from the intercom. Despite this racket, no one came to the door. Slightly concerned, I tried both bells again, to no avail. It was eleven o'clock and I didn't really want to press the doorbell again; I was aware that other people would probably now be cursing from their beds.

Casting around for options, I noticed a pub further down the road: Casey's. Warm golden light spilled out onto the pavement. On entering I was met by stony silence and staring locals. The only movement came from the hearth, where a log fire flickered. In a voice that sounded more confident than I felt, I asked after the owner of the B&B. I half expected him or her to be here, enjoying a pint of Guinness at the end of the day. The landlady took one look at me before helpfully suggesting I try ringing the doorbell.

I returned to the B & B, and while I was trying to work out what to do, still knocking and ringing at the door, I suddenly remembered that I'd scrawled down their phone number the night before. I could hear ringing inside as I held my phone to my ear. At least I had the right number. It rang eight times, then a shuffling sound and a muffled Irish accent: 'Hello?' 'Hi, I'm booked into your B&B for the night but no one seems to be answering the doorbell,' I replied, relief escaping into my voice in ripples. 'Oh right.' The shuffling sound again. 'I'm standing outside the door right now…' I hinted. A pause. 'Oh…' More shuffling. A further ten seconds passed.

Then the door swung open, releasing a blast of warmth and revealing one of the oldest men I'd ever seen. He was stooped over, a stained grey jacket hanging limply from his spindly frame. Wisps of white hair did little to cover his mottled scalp. He let me in. The air was stale and reeked of cigar-ettes. I glanced around the gloomy hallway, floral wall-paper decades old; a wooden desk with a white doily poking out from underneath a cascade of pamphlets; and a single musty armchair backed up into a corner under the stairs as if attempting to fade into the shadows.

The elderly man looked at me blankly so I explained that

I had booked a single room for two nights. Another few seconds went by. Then something seemed to dawn on his face and he led me upstairs, taking the steps at a sloth-like pace.

When we eventually reached the landing, he looked quizzically at me again and then, before I could stop him, he proceeded to throw open all the doors along the corridor. Suddenly alive, he crashed into each room, then shuffled out again. Some were clearly occupied. At last, he found my room. Saying goodnight, I made sure I turned the key in the door before going to bed.

Early Purple Orchids in their thousands welcomed me back to the Burren like an old friend. Two weeks older and they were showing their age. Pinks and purples were fading and many were crisping into brown spikes.

After breakfast, I'd taken a train north to Gort where I'd bought myself provisions for the day then headed west into the Burren. It was a beautiful day, in stark contrast to my last visit. The walk to Lough Bunny and the National Park was a pleasant ten kilometres along tiny roads bordered by rough limestone walls, pieced together like an intricate jigsaw puzzle and muffled with hibernating hedgehog-like mosses. Wet fields full of bogbean, a dainty white flower deserving of a prettier name, were soon succeeded by grassy meadows littered with large glacial erratics.

I passed Kilmacduagh monastery, a ruined abbey near Gort founded by St Colman mac Duagh in the seventh century. St Colman was a religious recluse and spent many years living and praying in a cave in the Burren. Beside the monastery is an Irish round tower, rising like an over-sized pencil from the graveyard. Bizarrely, the doorway to the tower must have been seven or eight metres above the ground. There were several backpackers with large rucksacks marching up the bumpy track to the ruins, covering their mouths and noses as passing cars sent up clouds of white dust. The round tower, supposedly the tallest in Ireland, leaned noticeably, as if burdened by centuries of tourists, pilgrims and exposure to the Irish elements.

In the distance, where the grey flags crumbled into silver rubble, there was a lone tumbledown stone cottage, a relatively rare sight in this part of the island. Throughout the western counties of Ireland, cottages like this stand deserted and crumbling, a hard reminder of the famine of the mid-nineteenth century, when funerals were commonplace and church bells rang across the countryside, mournful and morose. In the worst-affected towns and villages, where there was no one strong enough to dig individual graves, the dead were piled up and buried in large holes where the soil was loose and crumbly. Many died in their tiny cottages. Their homes would be their graves.

Such ruins are rare in the Burren, though, because the people here were so poor that they couldn't afford to build themselves houses using their own stone. While I and many others appreciate the wild beauty of the Burren, for many the place will remain forever tarnished by this dark period in Irish history. The poet Emily Lawless described the Burren Hills as 'skeletons – rain-worn, time-worn, wind-worn – starvation made visible, and embodied in a landscape'.

The sun was high in the sky now and I was beginning to burn. The countryside had lapsed into the familiar limestone pavement that I'd encountered two weeks previously, and there was very little shade. I had arranged to meet Sharon near Lough Bunny. While waiting for my lift, I began exploring the wet, tussocky grassland at the eastern end of the lough. A whitethroat began chirring from the top of a hawthorn. I stood still among the bright-yellow marsh marigolds and watched this little warbler sing its scratchy song. As I neared the lough, I spotted a single Early Marsh Orchid: pale magenta with a pinched lip of twinned loops and dots. The inflorescence was crowded with flowers, all desperately seeking the attention of passing pollinators.

The Early Marsh Orchid is so variable that it has been subcategorised into five different subspecies. Some, like the bright-red *coccinea* variety or the pure, white rarity *ochroleuca*, are easier to identify. Others, notably the one in front of me, are more difficult. It looked like the *pulchella* variety, derived from *pulchellus*, meaning 'beautiful' in Latin. But the Harraps suggested that this one is largely restricted to southern England. My dilemma didn't last for long, however; emerging from a thick clump of deep-jade reeds was another Early Marsh Orchid, this time with paler, flesh-coloured flowers decorated with squiggles of rose and flamingo pink: subspecies *incarnata*. The Early Marsh is quieter, more delicate and less showy than the other marsh orchids. Easily overlooked, it never forms the swarmed ranks so typical of its close relatives.

I wandered back along the road, occasionally jumping across to the limestone pavement to look for plants. Spring gentians still flecked the grass with bright-azure blue and the mountain-everlasting had burst into bloom, a scattering of clustered white flowers. Bloody crane's-bill poked out from the grykes, its dark-pink heads reminiscent of the marsh orchids. The air smelled sweet and summery.

I sat by the water's edge and gazed out over the lough, enjoying the light breeze and soft rippling of the water as gulls wheeled in the air above.

The Burren landscape has been inspiring poets for generations. W. B. Yeats, Seamus Heaney, Michael Longley and Seamus O'Sullivan have all written about these hills. In the constantly changing weather and shifting light, the Burren unfolds intimately as you pass through. Sudden showers, sharp bursts of sunshine and scurrying clouds transform the rugged landscape of limestone, ancient ruins, stone walls and meadows.

I was already experiencing its draw, after just a couple of days here. How can a place be so barren, yet so fully alive? So empty, yet bursting with interest? There is a sense of completeness; the kind experienced when all the components of a machine or a plan come together in perfect harmony. It clicks. The Burren is a memorial: it remembers the dead in its stone tombs; it remembers the water that has been slowly wearing it down over millennia; and it remembers the passage of poets, botanists and everyone else who has been inspired by its wilderness. Sitting there by the lough, I was filled with feelings too difficult to name.

A screech of tyres behind me brought me to my senses and I glanced over to see a grey Freelander that had pulled up at the side of the road. The window slid down and there was Sharon. She was in her forties, I guessed, wearing dark waterproof trousers, muddy from recent botanising, a khaki-green fleece and a hand lens looped around her neck. Now that it was obvious that Sharon was in fact a woman, I felt slightly ridiculous. Why had I thought anyone would name their son Sharon? Her voice, which had sounded so masculine on the phone, was clearly female.

I was so surprised and flustered that I tripped over my opening words, but Sharon was unfazed. We drove around Mullagh More to the same turlough I'd visited with Michael the week before. And, as we walked down to the shore, she explained that the Burren limestone was extremely flat and that Mullagh More, with its slanted layers of rock, was a rare exception. We climbed through a wall sprouting ferns and rue-leaved saxifrage and there, in front of us, was the very same grassy knoll where I'd spent half an hour searching fruitlessly for orchids only a week previously.

We split up and I quickly adopted plant-hunting mode, scanning the grass among the broken limestone fragments. And there, inexplicably, was a Dense-flowered Orchid. The search was over so quickly I'd barely had time to register it had started. The single plant, hiding amid the cottony tufts of the mountain-everlasting, was tiny. Its condensed head of creamy flowers were, as is typical of this species, all facing in the same direction. Each one sported a milky, dwarfed figure; smaller and more stunted than the Man and Lady Orchids, but a little person all the same.

What was particularly satisfying about this plant was that I had found it; to be shown it by someone else would have taken away most of the satisfaction, especially since I had already been to this very site. Annoyingly, there was an old spike next to it that had clearly been in flower on my earlier visit. I had been so close.

The Dense-flowered Orchid, or *Magairlín glas* to give it its Irish name, was a relatively late addition to our orchid flora. It was first discovered by Miss Frances More at Castle Taylor in May 1864, who reported gathering flowering spikes of a 'very ugly' orchid. She dried and pressed her specimens, sending some to the Natural History Museum in Dublin.

In *Life and Letters of Alexander Goodman More*, a biography of the naturalist A. G. More, Frances' brother, we learn how Frances and Alexander botanised the countryside surrounding Nut-wood and discovered immature spikes of 'an inconspicuous but evidently unfamiliar orchidaceous plant'. While Frances guarded the orchid, Alexander travelled to Dublin to meet Dr David Moore, director of the Royal Dublin Society's Botanic Garden at Glasnevin, to propose writing an Irish flora. He writes that 'the curious orchid now came into flower and my sister... collected and dried several specimens, remarking that the little orchis was something she had never seen before'. Between them, More and Moore decided that the plant was indeed the Mediterranean *Neotinea intacta* (a former name for *N. maculata*), and its addition to the British flora was announced by Dr Moore at the next meeting of the Royal Irish Academy. It has since been found across Ireland, mainly in the west of the country, in particular Clare and Galway.

Alexander More was the first of many botanists to express his surprise about its presence alongside alpine plants such as spring gentians and mountain avens, far removed from its Mediterranean heartland. Of all the mysteries surrounding the Burren's bizarre flora, the most puzzling is how these species all got here. The distribution pattern for *Neotinea* is highly discontinuous and unlike that of any other Irish species – except, More noted, *Arbutus unedo*, the strawberry tree. Could there be a link?

As it turns out, Irish fleabane, St Patrick's cabbage and the Kerry lily also show this restricted distribution split between the Iberian Peninsula and south-west Ireland. They are part of a small assemblage of plants known as the Lusitanian flora. It is a bio-geographical riddle. Sharon said one school of thought is that these Mediterranean plants migrated to Ireland after the last glaciation 12,000 years ago via wind-blown seeds. They could also be a relict flora left over from a time when Ireland's climate was more aligned with that of modern-day Spain. 'It's unlikely that *Neotinea* could have survived so far north during an ice age, though,' Sharon pointed out, 'particularly along with its symbiotic fungus and associated pollinators.' How and when the migration took place remains a mystery.

Some members of the Lusitanian flora, predominantly heathers, were shown to have been introduced to Ireland by the monasteries. This was a common practice more than a hundred years ago. Could monks have brought *Neotinea* over from Spain? And if so, for what purpose?

As Sharon and I walked around the turlough, I rang Michael back in England to tell him of the discovery. He erupted with laughter when I told him how close we'd been. We chatted as I wandered back through the limestone boulders, reminiscing about our time together. I updated him on my finds in Jersey and Kent. I was about to say goodbye when he stopped me as if he had something else to say. 'Oh and Sharon… is she a man or a woman?' I froze. Sharon was a step ahead of me and I was unsure whether or not she'd heard Michael's booming Geordie voice. 'Yep, she's been extremely helpful,' I countered. If Sharon had heard, she was doing a good job of pretending not to care.

In between the lemony shrubby cinquefoil and lilac heath dog-violets, we found more Early Marsh Orchids, but not the *incarnata* subspecies I'd seen earlier. These were the so-called 'leopard orchids', subspecies *cruenta*, that only grow at a handful of locations in Ireland and Scotland. The 'leopard' part was obvious: every part of the plant – leaf, stem, bract and even flower – was heavily flecked with dark-purple spots and blackened rings.

Through another wall we came across my thirteenth species of the year: a trio of Pugsley's Marsh Orchids. Their flowers were remarkably similar to the other marsh orchids, but, like the *Neotinea*, they all faced in the same direction. It was as if there was something particularly fascinating to look at. Up until now I'd thought orchids were all attention seekers, but here I realised that it was the reverse: no matter where I sat, many of them would have their backs turned away like stroppy teenagers.

Back in the car, Sharon told me more about the geology of the area. The shallow layer of limestone is vast, stretching all the way from the granite of Galway to the Cliffs of Moher and the Atlantic Ocean. The Burren didn't exist during the Carboniferous Period, 360 million years ago. In fact, nor did Ireland. Orchids hadn't evolved yet. During this time, a warm tropical sea covered the part of the world where Ireland now lies, and in it lived innumerable microscopic marine organisms with skeletons composed of calcium carbonate. When they died, they drifted down to the ocean floor and formed a calcareous mud. Over geological time, as layer upon layer formed, the mud became rock, the limestone we see today. Given how unstable the Earth's surface can be, the unmoved horizontal layers of limestone deposited more than 280 million years ago are testament to the stability of this area.

Despite our success with the Dense-flowered Orchid, Sharon seemed keen to show me some more of the Burren. She took me to Gortlecka Meadows, a collection of quiet grassy sanctuaries in the hazel forest. The Burren is famous for its dwarf hazel thickets, small pockets of spring that are the remnants of Neolithic coppicing. Three thousand years ago they would have provided pliable wattle for building fences and houses. These hazel spinneys have largely been abandoned for decades, left to reacquaint themselves with the wild.

Hazel was once a pagan tree, to be worshipped and treasured. To carry a hazel nut in your pocket was thought to be successful in preventing the onset of rheumatism, and the small buds, if eaten with dandelions, wood sorrel and chickweed, would cure colds and sore throats. It was also a May Day tradition to bring hazel twigs into the household to ward off evil spirits. Be warned, though, cut a branch of hazel for no reason and you would lose a year-old heifer.

We left the road and ducked onto a small path that disappeared enticingly into the hazel. It had been invisible just seconds before. Stepping through this miniature forest, with the canopy just above my head, felt eerie. The air was moist and mellow, the rocks mossy. Recent rain had formed a flash stream that had carved a channel deep into the path, depositing a chaotic flood of twigs and leaves. On either side, hazel branches hung low, heavy with spongy mosses and cupped lichens. They clung to the silvery bark, feathered arms drawing in moisture from the humid air. I've never been anywhere that felt so ancient. Centuries could pass and nothing would change.

We went through a gate made entirely from hazel, step-ping around lady ferns, male ferns and hart's-tongue ferns with their party-horn fronds. The bank ahead was a snow-drift of wild garlic. Suddenly something dog-sized and smelly exploded from behind a boulder with a shriek and scampered up the path. It was a goat. Sharon laughed as I wrinkled my nose at the smell. 'You probably interrupted its private time,' she joked.

Most of the goats in the Burren are no longer farmed and run wild, roaming from pavement to wood, through meadow and pasture. Sometimes they help keep the scrub at bay but more often than not their feeding habits do more damage than good. There is a heated debate among farmers about whether to manage the population, as they do with their own livestock so as to protect plant diversity. Traditional farming methods that have been in use for centuries are still maintained today. Many of these have been instrumental in forming the Burren as we know it. As ever, farming has the power to protect or destroy what's special about the land it is practised on.

At the bottom of the hill, we hopped over a gate and out of the hazel thicket. Gortlecka Meadows sloped smoothly away from the trees, gentle and unassuming. It was yellow, pink and vibrant green.

We began searching in the grass while discussing my species tally to date. Sharon was impressed, and particu-larly jealous of my trip to Jersey. We had been walking for several minutes, through the most amazing array of spring wildflowers, when Sharon suddenly stopped. 'Have you got Fly Orchids yet?' she asked. I paused, the beginnings of my answer disappearing in the wind, as there in front of us were two freshly flowering Fly Orchids, side by side in the middle of the meadow.

Standing tall, and yet incredibly difficult to spot, their spindly stems carried the first of many camouflaged flowers. Like the Early Spiders I'd found on the Dorset coast, the Fly Orchid is one of the *Ophrys* orchids: the insect mimics. It has two wiry antennae that perch on its flower, surrounded by brilliant lime-green sepals. Its velvet body tends to be narrow and chocolate brown, with a silvery-blue band across its middle that imitates the sheen of an insect's wings. It often grows in woods where little direct sunlight reaches the floor. Here at Gortlecka, though, the plants were right out in the open.

Just as the Early Spider isn't visited by spiders, the Fly isn't pollinated by flies. *Argogorytes mystaceus* is a digger wasp: thin, with an hourglass figure and a black body syncopated with narrow bands of yellow and gold. It looks very different from the Fly Orchid flower. So different, in fact, that you really start to question the wasp's intelligence. How can it possibly be duped into thinking the orchid flower, whose appearance is only marginally insect-like, is its star-crossed lover? It turns out the flowers have gone for a slightly different strategy. They've paid less attention to how they look and invested more in developing the way they smell and feel. The scent, undetectable by the human nose, would be musky and meaty, a carbon copy of the pheromones emitted by the female. Each lip is covered in soft, silky hairs. To the digger wasp, the Fly Orchid is a sex toy, not perfectly life-like but able to arouse the senses and cloy the mind. It sates their sexual desires while they wait for the females to hatch, and in return they pollinate the flowers. Tit-for-tat.

It was after only a few minutes of sitting and watching that I realised we were being steadily surrounded by Fly Orchids, true masters of stealth and camouflage. They appear slowly and softly, shifting in and out of focus. The longer you wait, the closer they get. You'll see one three metres away, yet remain unaware that one has crept right up to your knee. Over the years, I've realised that looking for Fly Orchids is a futile activity; their ability to vanish right in front of your eyes is unprecedented. Instead, you have to let them come to you.

With this success bringing me up to fourteen for the year, we continued to explore the Burren. Driving up the winding lanes, round corners so sharp they could only have been designed to keep tourist coaches away, I wondered what it would have been like to live here three or four thousand years ago.

Evidence of past inhabitants can be found scattered all over the Burren. 'It's equally fascinating from an archaeological point of view as a botanical one,' remarked Sharon, 'and you'd be surprised how often new sites are discovered, dug into pockets of the limestone pavement.' You can find evidence of Stone Age, Bronze Age and Medieval settlements in close proximity to one another. The sheer number of burial sites is astounding.

We passed a Neolithic wedge tomb at the top of the hill, one of eighty in the Burren. A flat roof stone slanted across several upright boulders. I marvelled at the strength it must have taken to place it there. What kind of person had been buried below? A farmer? A hunter-gatherer? Had they been awestruck by the rocky landscape or simply taken it for granted, living their everyday life until it was finally over and they were left here, buried on the hillside? Perhaps they'd hated it, repulsed by bleak associations with hunger and hardship. Sharon pointed out that it was more likely the grave of several people. She gestured at another tomb on the horizon which, she said, dated back almost 4000 years. It was a burial site for six children and as many as twenty-two adults. Their remains had been found with numerous personal items: stone tools and simple jewellery.

We had reached the edge of the limestone pavement where the remaining patches of silver stone petered out through the grassland, like the last bits of snow that refuse to melt. I knelt down in the warm grass to admire a clump of twelve Dense-flowered Orchids that Sharon had come across the previous day. They seemed dangerously close to the road, constantly threatened by the swerve of a wheel. So small and inconspicuous, I couldn't really blame myself for not finding them the week before. Each little floret was like a tiny vanilla ice cream laced with lemon and lime. Why do you choose to live here, I thought to myself, when you could be basking in the Mediterranean?

Further up the road, I found a few plants that had duller flowers, a band of pink running up the centre and pock-marked ovaries. These unhealthier-looking plants were of the variety *maculata*, which is much rarer in Ireland than the creamy-flowered specimens I'd seen already, but by far the commoner type in Spain and Italy.

I sat and enjoyed the sunshine, watching the beautiful white and roan of the shorthorn cattle as Sharon wandered further down the lane. I'd had a wonderful time in the Burren and felt privileged to have been there. I counted five more Dense-flowered Orchids – the creamy kind – marching in single file through the limestone. The juxtaposition of the vanilla flowers, pure and delicate, and the jagged clints; of fragility and extreme endurance, was humbling beyond comparison. I leaned back, soaking up the wilderness of the Burren as it tightened its hold on me, ensuring I would return one day, with so much more to explore.

6

Swords of the Hampshire Hangers

'As a quoted line of verse will suddenly evoke by association a complete poem, so a specimen of any given plant will call up not only all the flowers associated with its particular habitat, but the whole feeling and atmosphere of the place itself.'

Jocelyn Brooke, *The Wild Orchids of Britain* (1950)

Hampshire
June 2013

Steve Povey started botanising at the age of six. His father was a keen naturalist and would frequently take him out into the Chilterns to look for rare plants and insects. On one of these forays, he distinctly remembers seeing his first Bee Orchid, which he fondly recalled 'appearing enormous to a little lad!'. Slightly put out that he'd found his first Bee Orchid at a younger age than I had, I tried not to make things a competition; I got the impression I would lose. Later, around the age of ten, he was shipped off to his older cousin in Sussex for every school holiday. She was an avid botanist and particularly loved the chalk grassland of the South Downs, a habitat that remains Steve's favourite to this day.

June had begun bright and clear, the sun beating down in an effort to make up for lost time. I drove to Petersfield to meet Steve, who had agreed to join me for a day of orchid hunting in Hampshire. As I walked across the car park to his battered silver Ford Mondeo, I knew it would be an exciting day. Dangling in the front window, hung from the rear-view mirror, was a cardboard cut-out Sword-leaved Helleborine.

As he drove me along Hampshire's winding country lanes, we talked about the challenge I'd set myself and its progress so far. He asked me to list the species I had never seen before and kept interrupting to suggest sites I should visit. We both agreed that the Small White Orchid would be one of the hardest species to find.

Today, we were off to see Britain's largest population of Sword-leaved Helleborines, now one of Britain's most vulnerable orchids. Despite this, it would also be one of the easiest to find: 85 percent of the British population's several thousand plants, grow in a Hampshire woodland, not far from Petersfield. The only time I'd seen this orchid before was when I was sixteen. I'd managed to drag my sisters along for the day and persuaded my mother to let me take a break from revising for my GCSEs and drive us on the eighty-mile round trip to Hampshire.

The genus *Cephalanthera* is an older, more primitive group of helleborines and includes three UK species: Sword-leaved, White and Red Helleborines. While the first two are scarce but widespread, the Red Helleborine has an extremely restricted distribution, with only two or three populations. The name *Cephalanthera* comes from the Greek *kephale*, meaning 'head', and *antheros* or 'flowery', which together describe the large inflorescence.

In *Wild Orchids of Britain*, Summerhayes fondly recalls how 'the contrast between the white flowers and the fresh green leaves produces a charming effect, which is much enhanced if, as is sometimes the case, there are several plants growing together'. Steve's first encounter with Sword-leaved Helleborines was just as Summerhayes described.

In the late 1970s, he was asked to survey a large garden just north of Petersfield which was due for development to see if there were any ornamental plants worth saving before the bulldozers moved in. As he turned into the property, he noticed the long rose border that lined the driveway. It wasn't the roses that drew his attention, but several dozen Sword-leaved Helleborines. Unable to believe his eyes, he didn't recognise the orchid in front of him for what it was straight away. He'd never seen Sword-leaved Helleborines before, let alone in the rose border of a semi-suburban garden. He scrambled out of the car for a closer look at this fine colony of *Cephalanthera longifolia*, growing in the most unlikely of places.

Steve had recalled that this species was more common just north of Petersfield than anywhere else in the country, but surely that didn't mean looking for them over garden gates. He tried to save some of these plants, but as with many orchid species, they didn't enjoy being moved around so unfortunately none of them survived. Now living near Petersfield, Steve sees these orchids all the time and is a regular visitor to the copse we were heading for. If we were lucky, he said, we might find the rare hybrid of Sword-leaved and White Helleborines, *Cephalanthera* x *schulzei*.

We arrived at the copse a little before midday with the sun high in the sky and slipped blissfully into the shade of the beech trees. Bounded by agriculture, this small patch of woodland is a sliver of tranquillity in the surrounding monotony of arable fields, a delightful everyday miracle. Tall, ivy-strewn beeches hold up the golden-green ceiling, criss-crossed by a fine skeleton of branches. The leaves took the bite out of the sun, leaving behind the cool warmth of a mid-summer's day.

Within ten seconds of going through the gate, I had already singled out my first Sword-leaved Helleborine. Or rather Helleborines, because the woodland was carpeted with their graceful white spikes. They are majestic plants; tall, with two ranks of narrow, pointed leaves. Its other name, Narrow-leaved Helleborine, is far too mean for a plant like this. The flowers are pure white and balance graciously atop the fresh green leaves, providing a stunning contrast. The top of the lip has intricate parallel ridges covered in a dusting of yellowy-orange pseudopollen – pretend pollen to lure pollinators to the flower without actually providing them with anything. They take, but don't give. To me, it looked as if each flower had its mouth wide open, revealing an orange tongue and a row of sharp, pointed teeth. Unnervingly, there were no eyes to accompany it.

Steve's friend Jeff Hodgson was there when we arrived. A tall, friendly Yorkshireman in his sixties, he had come down from north Wales for a few days to complete an orchid tour of the south. He had been down to Kent the previous day looking at Lady Orchids. Even though I'd seen them so recently, I felt a moment of envy. This summer of orchid hunting was already making me greedy.

Jeff led us further into the wood, soon turning off the main path and following a small badger trail that ran between the trees. There were Sword-leaved Helleborines everywhere. The population here is not only the largest in the country, but also one of very few that are actually increasing in size. Most colonies have fewer than fifteen plants and are therefore very vulnerable. Even here a large storm could have a severe impact on this population. While selective tree felling is used in the conservation of this species, random removal of large numbers of beech trees in a storm would allow too much light to reach the woodland floor, which would quickly be taken over by asphyxiating bramble thickets.

Sword-leaved Helleborines were first discovered in Helks Wood, near Ingleborough, in 1666. This location was already known for its orchids, as it was the place where the Lady's Slipper had first been recorded several decades previously. Despite being considerably rarer than its cousin the White Helleborine, *Cephalanthera damasonium*, the Swords are more widespread. Even today it can be found in Wales, Scotland, Ireland and in central and northern England, although in tiny numbers. The Hampshire hangers, though, always have been and probably always will be the Sword-leaved Helleborine's heartland.

After a minute or so of walking, Jeff suddenly pointed and rushed forward, bending down to look at what initially appeared to be two Sword-leaved Helleborines in bud. Steve knelt to examine the plants, checking bracts, upper leaves and sepals; no characteristic went unscrutinised, no detail unnoticed. He liked what he saw, rapidly arriving at the conclusion that these were the *schulzei* hybrid. The Sword-leaved Helleborines were in full flower all around us but all the White Helleborines were still tightly enclosed in their buds. These hybrid plants were evidently just about to burst into flower. The leaves were also different, not nearly as long and narrow as in *longifolia*, but clearly pointier and less broad than the *damasonium's*. Steve showed us several smaller, but important differences that made this plant so transitional in its appearance, confirming beyond doubt that it was the rare hybrid.

I felt myself becoming too bogged down with which plants were hybrids and which weren't, so wandered off further into the wood to clear my mind. Within seconds, I had found my second new species of the day. Under a large beech, a clump of Bird's-nest Orchids: my sweet sixteenth.

The Bird's-nest Orchid is one of the weirdest plants I've ever seen. Completely brown, it appears at first glance to be dead, but a closer examination proves otherwise. Each flower is velvety caramel and has two feet that look as if they've been drawn by children: big, clumsy and sticking out sideways. Some plants were still in bud, looking like bizarre trees covered with peanuts. This orchid never produces chlorophyll – the green pigment used in photosynthesis to help produce sugars – a fact that makes you question what qualifies as a plant.

Unlike most species, the Bird's-nest Orchid's name isn't derived from the appearance of its flowers or leaves, but from its roots, which form an entangled mess superficially similar to a bird's nest. It was first recorded in 1597 by John Gerard, who rather rudely calls it a 'bastard or unkindely satyrion' (satyrion being the name for an orchid at the time). He does, however, describe it wonderfully as having 'many tangling roots platted or crossed one over another very intricately, which resembleth a Crowes nest made of stickes; from which riseth up a thicke soft stalk of a browne colour, set with small short leaves of the colour of a dry oaken leafe that hath lien under the tree all the winter long'.

German botanist Hieronymus Bock, writing even earlier than Gerard, noted that it grew in the woods and hedge-rows where birds nest. He decided that these orchids must grow where the semen of small birds has fallen to the ground. Indeed, this concept was extended to other species too: that orchid flowers rose up from the semen of the animals they resembled. This idea survived for more than a century.

So how is it that a plant with no chlorophyll, and there-fore no means of capturing energy from the sun, can survive and grow? It is often written that Bird's-nest Orchids are saprophytic, obtaining their nutrients from decaying leaf litter. However, this is not true. Instead, the orchid makes use of the mutualistic relationship between a fungus of the genus *Sebacina* and trees such as beech. The tree photosynthesises and passes carbohydrates on to the fungus, which in return supplies the tree with vital mineral nutrients that are other-wise difficult to obtain from the soil. The Bird's-nest Orchid cheats by invading this system, digesting the fungus and in doing so obtaining carbohydrates indirectly from the tree, offering nothing to either the tree nor the fungus in return. In this way 'parasitic' would be a more accurate adjective than 'saprophytic'. One end of the fungus is attached to the tree, receiving carbon produced by photosynthesis; the other end is attached to the orchid, which is siphoning off this carbon. They are outlaws, sneaky thieves who execute their criminal-ity with perfection.

Because they aren't dependent on light for food, Bird's-nest Orchids are almost always found in dense shade, often in the absence of other herbaceous plants. The ones in Hampshire were no exception. The woodruff had backed off, leaving the orchids all alone in the beech leaf humus. Being very careful not to squash any of the surrounding spikes, I knelt down on the ground and examined one of the larger brown plants. It looked as if it'd been coated in golden honey. Where the light filtered through the canopy, the orchids were lit up ginger and orange.

They really are exceptional when you stop and think about it; the overwhelming majority of the world's 400,000 species of land plant manufacture their own food using chlorophyll, so how is it that this tiny orchid has evolved to take a risk and abandon that strategy, potentially putting its very existence in the hands of an organism as tiny as a fungus? In this small Hampshire copse, where the woodland floor was home to sweet woodruff, bugle and Sword-leaved Helleborines, these Bird's-nest Orchids were definitely the odd ones out.

I returned to where Jeff and Steve were still admiring the hybrid helleborines before we made our way back to the main track. Steve began telling me about his adventures with the late Francis Rose, an enthusiastic botanist whose knowledge of the British flora was unrivalled. Rose had never been able to drive, and his wife no longer wanted to travel for long distances so Steve and his friend had started to take him orchid hunting on a regular basis.

Steve first heard of Francis Rose when he moved to Selborne in the late 1960s. He lived in the next village and was a close friend of Lady Anne Brewis, who was the BSBI county recorder at the time. Steve regularly visited the nearby Noar

Hill Nature Reserve, an area of late-Medieval chalk workings that boast a fine array of butterfly and orchid species. He recalled how he would often see Rose walking about the reserve either alone or with a colleague. 'As with most such people you are to some extent in awe of them,' he said. 'It took a while before I plucked up the courage to approach him.'

One day, while on Noar Hill, Steve discovered what he thought was a hybrid of Chalk Fragrant and Common Spotted Orchids. Rose happened to pass by while he was studying the plant so he asked if he would offer his opinion. He came over, took one look at the plant and then, with a smile, happily agreed with Steve's identification. He then asked if Steve would like to help him find some unusual Bee Orchids that he had spotted on the reserve the previous year, and so the ice was broken and a friendship began.

Steve remembers him as very calm and pleasant, but also very resolute. On many occasions, when Steve was planning a trip to some out-of-the-way place in the hope of finding a particular species, he would pop round to the Roses' house armed with a map of the area in question. In a matter of moments, Rose was not only able to tell him precisely where to find his plant, but also what else he could expect to see while he was there. He'd always say: 'You must also go here and here, oh and this place is fantastic for this', scattering the map with crosses. He seemed to be able to do this for just about anywhere in the country.

Francis Rose sounded like a remarkable companion and Steve told me how he wished he'd had a tape recorder with him, for during each journey to a particular site he would recount in great detail many of his own memorable experiences of discovering particular orchids after many years of searching, and occasionally he would describe the rivalry that existed between himself and what are now considered to be some of our finest early twentieth-century orchid hunters. These tales were legendary, and not recording them remains one of Steve's greatest regrets.

On a number of occasions, they visited sites where many years before Rose had found some rare species of orchid and thought it would be good to see if they were still there. Steve laughed as he told me how 'it was always surprising, even in his later years when walking was becoming less easy, how much time he would dedicate to refinding them – far more than I would probably have given. Off he would go, fighting his way through the thickest undergrowth, often out of sight for an age before appearing again, slightly dishevelled and often with twigs and leaves caught in his hair. But with nothing to report he would go off again. In a few memorable cases his persistence paid off, when almost at the last minute, just before deciding it was a lost cause, he'd discover a few plants'.

I was particularly jealous of one trip to Kent: 'On one occasion in late June he showed us a secret site for Late Spider Orchid; there were about twenty plants in a group, each with around twelve to fourteen flowers. I have never seen anything like it, and only about thirty yards from a main road!'

Reaching a fork in the path, we came across a loose collection of sticks stuck in the ground on either side of the track. These I remembered from my visit three years previously. At any famous site, wherever it is in the country, there will always be regular enthusiasts who visit throughout the flowering season. When they find new spikes coming up or in flower that are vulnerable to damage from the path, they will insert a twig into the ground to warn people of their presence. Of course, they also act as a guide to orchid hunters visiting the site for the first time.

This particular group of sticks were marking Fly Orchids. It seemed a long time since I'd seen the two plants in Gortlecka Meadows in Ireland. There were several spikes in flower here, all bearing only one or two small conker-brown flowers. The chequered light fell upon the low-lying and sparse ground vegetation, providing the perfect habitat for this small orchid. It's one of my favourites, quietly going about its business. They were easy to miss in the dappled beech-brown shade.

Tree felling in woodland is a serious threat to the Fly Orchid. One would not think that this species would be in danger from over-collecting, given its somewhat dull appearance. However, there are stories of orchid hunts that ravaged populations. In 1898, S. L. Petty reported how collectors had raided the local woods for Fly Orchids: 'dozens of people with baskets (and sometimes trowels too) invaded the woods, and, of course, asked no permission to take roots away, but did so.'

Back in the car, Steve and I headed further into Hampshire's beech wood hangers. He talked of finding Ghost Orchids in Herefordshire, a total of seven or eight in his lifetime. Seven or eight! I would be indescribably lucky to see just one. He spoke of his love for the British flora. He claims to have seen about 95 per cent of our country's plants, which is an

incredible feat, and said that, if he wanted to, he could probably squeeze out another two or three per cent before he
stopped travelling around. I was trying hard not to sound too
jealous.

We drove for half an hour along a rollercoaster road,
through the hangers, the cardboard Sword-leaved Helleborine swinging madly to and fro. The word 'hanger' derives
from the Old English *hangra*, meaning 'wooded slope'. Typically, beech hangers are extremely steep and are home to an
impressive orchid flora throughout the season, if you know
where to look.

The Hampshire hangers have historical connections with
the eighteenth-century naturalist Gilbert White, who used to
walk in the beech woods above Selborne, very near to Steve's
house. Selborne is a charming little village full of thatched
red-brick cottages and gardens overflowing with roses and
wisteria. It sits at the base of the thickly wooded hanger
immortalised by White in his famous work *The Natural
History of Selborne*. In his letters, he writes of his fondness for
the beech woods: 'The covert of this eminence is altogether
beech, the most lovely of all forest trees, whether we consider
its smooth rind or bark, its glossy foliage, or graceful pendulous boughs.' Because he was an ornithologist rather than
a botanist, though, very little of White's letters refer to the
plant life growing in the hangers.

Steve was taking me to another population of Sword-
leaved Helleborines. He assured me that not many people
knew of this colony, which grows halfway up the hanger.
Compared to the copse we'd visited earlier, this wood felt
nearly vertical. Trees grew out from the scarp slope at absurd
angles. The chalky path zigzagged vertiginously upwards. I

found myself eye-to-leaf with some of the tallest beech trees I've ever seen. The world had been tilted on its side.

Botanically, the hangers are arguably the richest of the chalk woodlands in England. We waded through some ramsons and the strong smell of garlic wafted along with us. Edward Step, in *Wild Flowers in their Natural Haunts*, writes how their 'flat-topped umbels of white star-like flowers' bloom in the woods in spring. He goes on to say that 'a few weeks ago many persons would be mistaking its broad oval leaves for those of Lily of the Valley, until they happened to tread on one, and then the unmistakeable odour of garlic would quickly undeceive them. The flowers gathered in igno-rance of the species are usually soon thrown away for the same reason – their broken stems give off too strong an odour to be pleasant'.

We paused briefly at a three-way fork while Steve tried to remember which way to go. Yellow archangel and bugle vied for attention in front of the beech and yews hanging precar-iously onto the slope. The weird cup-shaped flowers of wood spurge glowed lime green on their gangly stems.

Steve decided we should take the left-hand path. The scarp slope plunged downwards on my left as we walked along the chalky paths. It was a dizzying drop down into the valley. This was Edward Thomas country, and I wondered whether he had ever found *Cephalanthera* in the hangers above his home in the aptly named village of Steep. Had he been inspired by the pure-white blooms? Did he ever compose any poems about this orchid? I liked to think so.

Shortly we arrived at a small clearing carpeted in white: Sword-leaved Helleborines. I hadn't expected to find them growing out in the open, and yet here they were, spreading out into the clearing, completely exposed yet evidently thriving. It was a wonderful sight: swathes of bladed bottle-brush spikes with the Hampshire hangers visible through a gap in the trees behind them. There must have been more than 200 plants scattered through the grass, and Steve assured me that this was almost definitely the second-largest population in the country. And yet barely anyone knows about it. He told me that a recent nationwide survey listed all the sites and the number of plants found there, and this place wasn't even mentioned.

I felt there was more to this visit than met the eye. There was a great sense of inheritance about it. It was as if this was Steve's population and he was entrusting the knowledge of it to me so that I would be able to visit and look after it when he was no longer able to. As if reading my mind, Steve recalled something his father had told him when he was a child: 'If you ever gain any knowledge, make sure you pass it on while you're still around.' A secret colony of a rare orchid is a precious thing, particularly one of this size, and I felt honoured to have been entrusted with it.

The Sword-leaved Helleborine is so tied into my experience of steep beech hangers and Hampshire woodland that just seeing photos of it evokes memories of that warm summer day. Jocelyn Brooke found that 'as a quoted line of verse will suddenly evoke by association a complete poem, so a specimen of any given plant will call up not only all the flowers associated with its particular habitat, but the whole feeling and atmosphere of the place itself'. I knew I would never forget the Swords of the Hampshire hangers scattered beneath the towering beech trees, their pure-white flowers twinkling in the gloom.

7

The Desirable Category of Very Rare Orchids

'It was seldom, in the social milieu frequented by my family, that I encountered anything so exotic and orchidaceous as Miss Trumpett.'

Jocelyn Brooke, *The Military Orchid* (1948)

Oxfordshire and Buckinghamshire
June 2013

Miss Trumpett, in the words of Jocelyn Brooke, was 'full-lipped, with powdered cheeks of a peculiarly thick, granular texture, and raven-black frizzy hair'. She would regularly spark gossip among the village curtain twitchers, appearing on summer afternoons dressed in resplendent gowns 'worthy of Ascot', complete with extravagant, colourful hats. Brooke was utterly infatuated with her, fascinated by her clothes, her rich, resonant voice and her Latin American roots; it does not seem unreasonable to imagine that he spent many a botanical

expedition daydreaming about her. But after a single romantic liaison and the accompanying disapproval of his nurse,
Brooke eventually decides that perhaps they are too different,
her world too alien, for the romance to ever fully flourish.

The modern-day association between orchids and the
exotic is epitomised by Brooke's depiction of Miss Trumpett
as orchidaceous. The two words have become so closely linked
that I am often met with amazement when I say that there
are more than fifty different orchid species growing wild here
in temperate old Britain; people assume that orchids can
only be found growing in the decadent depths of the Amazonian rainforest – or in the local garden centre. But while
our islands may not host anything on the same scale as the
Cattleyas orchid and *Phalaenopsis* orchid that abound in our
supermarkets, a quick flick through the centre pages of this
book will hopefully convince you that the variety of forms
our native species take are anything but bland. They can be
just as alluring to collectors as the showiest of *Paphiopedilums*.

During the nineteenth century, when Orchidelirium was
sweeping across the country, many who could not afford the
steep prices for extravagant bouquets of Amazonian orchids
simply went local; they took to the woods and downs, plun-
dering populations of Ladies, Bees and Bird's-nests and
returning home with baskets laden with specimens. Some
would be sold by the bunch at local markets, while others
would end up in the vast hervaria, consisting of a multitude of species and varieties, assembled by naturalists. The
legacy of this Victorian obsession with plant collecting is still
evident today, when we consider the plight of so many of our
native species.

One plant that suffered particularly from the depredation

by nineteenth-century collectors was the Monkey Orchid. In his book *Wild Orchids of Britain*, V. S. Summerhayes writes that 'being one of our rarest orchids, and [with] the remarkable resemblance of each flower to a small monkey, this species has always had a special fascination for orchid lovers'. He goes on to note that 'the species was until about 1835 still widely distributed and locally plentiful on both sides of the Thames' but 'within a comparatively short period of time it had become rare'. He suspects that 'rapacious collecting at one period probably contributed to its disappearance'.

The Monkey Orchid was first recorded in Britain in 1666 by Christopher Merrett, who found it alongside the Military Orchid 'on several Chalkey hills neer the highway from Wallingford to Redding on Barkshire side of the river'. While old records for this species can be dubious, given the tendency people had to confuse the Monkey with Military and Lady Orchids, it is clear that it was once quite frequent in the Chilterns. The species then went into rapid decline, partly due to voracious collectors but also as a result of the ploughing of downland and the increase in the local rabbit population. Within ninety years, the number of significant Chiltern populations had dwindled to a single site near Goring-on-Thames called Hartslock.

During the 1920s, the population at Hartslock remained stable, with over a hundred flowering plants, and continued to increase through subsequent years. Unfortunately, the site took a hit in the late 1940s, when the lower meadows were ploughed up. After the Second World War, the Americans had sold off their surplus tracked machinery to local farmers who, now equipped with new firepower, proceeded to churn up the local downland in order to combat the national food shortage. Fortunately, the steepest part of the slope remained untouched and a few plants were able to survive here; they still grow there today. After a painful few years, during which the number of flowering plants remained in single figures, Hartslock was bought by the Berks, Bucks & Oxon Wildlife Trust (BBOWT) in 1975. Despite introducing new management strategies, it took a while for the orchids to get going again but the number of plants has steadily increased and since the turn of the century the site has supported 300 plants, of which at least one-third flower each year.

And Hartslock is no longer Britain's only site for the Monkey Orchid. The Harraps tell the following story of its discovery in west Kent in 1952: 'a single Monkey Orchid appeared on the rough grass of a disused tennis court at a vicarage at Otford. Every year, until 1955, the vicar took the seed capsules and scattered seed onto nearby downland. In 1956 there was a single robust spike and a further six non-flowering plants at the vicarage. However, on the retirement of the botanically minded vicar the new incumbent would not guarantee to safeguard the colony. All the orchids were moved to nearby private land, where the largest flowered once only, in 1957, and then the Monkeys vanished.'

The Monkey Orchid had in fact been recorded near Faversham in Kent in 1777. A few plants in the area over the next two centuries, and in 1955, the first of a string of new plants appeared near the town, where they continue to grow to this day. Between 1958 and 1985, the plants were hand pollinated to ensure seed was produced and numbers gradually increased, with more than 200 flowering spikes in 1965. Today the site remains private but still supports similar numbers. In the same year that hand pollination was instigated at Faversham, seed was taken and scattered at other sites in Kent in the hope that further populations would establish themselves. Most failed, but by a stroke of luck, the Monkey Orchid took a liking to Park Gate Down, where the population has since stabilised at about a hundred plants and continues to increase. This is the third and final known site for Monkey Orchids in Britain.

Interestingly, the Kentish Monkeys differ noticeably from the Chiltern plants. They are on average taller and more robust, with bigger leaves and larger purple spots on the lip, and are much more richly coloured than their Oxfordshire relatives. At the same time, there is a lot of overlap between the populations and it is thought that any variation is well within the normal range for this species. Morphometric work by Bateman and Farrington showed that morphological differences between the two are negligible. Considering the two as separate varieties would be like taking two daisies from your front lawn, one with pinkish petals and the other pure white, and assigning them and their progeny different names, despite the fact that they both fit within the average range of colour variation within the large population on the lawn. While there is very little variation within populations, particularly in Oxfordshire, this is thought to be a reflection of a highly depleted gene pool as a result of passing through a genetic bottleneck.[1] Given that the colonies are so isolated, genetic deterioration is a real threat.

---

[1] A genetic bottleneck occurs when environmental conditions result in a rapid decrease in population size to a very small number of individuals. The population recovers, but with a much-reduced gene pool.

A few days after my visit to the Hampshire hangers I drove up to Oxfordshire to visit the Chiltern Monkey Orchid population at Hartslock. Red kites circled above the A34 as I headed north, clearly enjoying the rising thermals and increasing in frequency the further I went. It is easy to judge how near you are to the Chilterns by counting the number of red kites in the sky; five or six a minute and you are nearly there. I had been to Hartslock once before, eagerly dragging my parents out one Sunday afternoon in 2009, three weeks earlier in the year. My notes from that day are barely legible, a reflection of my excitement, the occasional awestruck sentence scribbled down in between photographing Monkey Orchids.

I got to Goring late morning and parked in a small layby at the bottom of the narrow winding lane. The sun was already high in the sky, turning the road surface ahead of me into a nebulous haze: it was going to be a scorcher. The reserve is named after a lock on the river owned by the Hart family in the early 1500s. The lock was demolished in 1910, but the family name lives on up on the hill.

I left the road and began climbing up through the woods, woodruff and sanicle succeeding rough chervil and common vetch as I went. The trees held the fresh green of spring, and sunshine streamed in through the leaves dappling the woodland floor. A cluster of ghost-pale White Helleborines grew at the top of the hill under a row of large beech trees, providing a surprise seventeenth species before I had even reached the main slopes.

While similar to the Sword-leaved Helleborines I had

seen in Hampshire, *Cephalanthera damasonium* is far less extravagant, its flowers being more dumpy and off-white – earning it the name Egg Orchid in parts of the country – not quite attaining the purity of its rarer cousin. Not only this, but they don't open nearly as widely. In fact, the flowers of the White Helleborine barely open at all, often leading unknowing admirers to believe they are still in bud. Inside the flower, at the base of the lip, are three golden-yellow ridges much as in *longifolia*, which supposedly guide insects such as bees into the mouth of the flower, where they come into contact with the mass of yellow vanilla-tasting pseudopollen. Edward Step, writing about June in the woods in his book *Wild Flowers in their Natural Haunts*, recalls finding a White Helleborine in the shadow of a beech, with which it is associated.

'The White Helleborine is a wasp-flower: therefore it has no hollow spur for the secretion of honey, for the wasp has no long tongue like the butterflies and some of the bees... One is reminded, by this difference in the structure of different Orchids to suit the mouths of special insects, of the fable, attributed to Aesop, of the scurvy tricks played by Fox and Stork upon each other. The Fox invited the Stork to a feast, but the wines were served in shallow dishes from which the Fox could lap them but the Stork could not. Then the Stork, in returning the compliment, had her dainties served up in long-necked vessels in which she could insert her long beak, whilst the Fox had to be content with licking off the outside what little chanced to overflow. The Orchids, apparently, are not actuated by mischievous motives of that sort, but provide suitable accommodation for the guests they desire.'

The White Helleborine is a somewhat introverted orchid. Sarah Raven describes it as a 'straight-laced librarian' of a flower, a 'spinster who turns herself out neatly in public', sentiments echoed by Jocelyn Brooke, who writes that 'owing to the unwillingness of its flowers to open, the White Helleborine has an oddly self-absorbed, unforthcoming appearance'. But Brooke also talks of the handsome spectacle of a White Helleborine in full flower when it resembles 'on a small scale some exotic orchid from the tropics which has somehow contrived to stray into an English beech wood'. It enjoys a quiet existence under the grandest of beech trees.

Despite their shy nature, I hold a special affection for White Helleborines. During the summer of 2011, eleven years after moving into our house, a scattering of *damasonium* appeared under the beech tree at the bottom of the garden. I had initially dismissed the first one as an outlying lily-of-the-valley from a nearby flowerbed, never for a minute believing that this Cephalantheresque plant was actually a helleborine, only to discover that, as May moved into June, my preliminary hunch had been correct. Brooke mentions that 'it has also been called the Lily-of-the-Valley Orchid, though the reason for this is not apparent'. I felt surprisingly smug upon reading this statement, for I knew exactly why it had been given this name. That year I counted more than fifty plants both under our tree and in the woods over the road. They flowered again in 2012 and had sent up more shoots that year, although they remained tightly in bud on the front lawn. The number of flowering plants has steadily diminished since that first discovery, but they still come up every year without fail, usually producing two or three large spikes among a scattering of smaller ones.

I emerged from the wood at the top of the hill, blinking slightly at the strength of the sun. The reserve looks down over the Thames as it meanders its way towards London, the water peacefully slipping by just as it had done more than 500 years ago when the Hart family installed their lock. I was surrounded by a panorama of softly undulating hills, one wooded slope rolling into the next. It was idyllic, so you can imagine my disappointment when the view became some-what ruined by a thick white tape spread haphazardly across the hillside. Each year the warden cordons off the main group of Monkeys, arranging little pathways so that visiting orchid enthusiasts can still gain access to the best plants for pho-tography. It is a necessary precaution, don't get me wrong, but it detracts so much from one's encounter with the orchids.

As I walked down one of the paths, so well used by this point that there was no grass left at all, I saw the first Monkey Orchids, standing timidly among the browned cowslips. There are no words to do justice to the extraordinary wack-iness of these floral aristocrats, for aristocrats they are. The flowers, which crescendo open from the top down rather than the usual bottom to top, are white and pale purple in colour; each lip is moulded into a little monkey-like figure with a looping tail and long, flexible limbs. Some danced for joy at the sight of other Monkeys on the hill; others longed to race through the jungle of plants in their hillside meadow.

How or why they evolved is a question that still baffles even the most eminent of evolutionary biologists. What pos-sible advantage could there be to having a monkey-shaped flower? Perhaps it's merely a reduced landing pad for visiting insects, the curling appendages acting as runway lighting, but somehow that spoils the magic of it.

Brooke, like so many others, found the Monkey Orchid enthralling. He, too, acknowledges its aristocratic nature, 'for about certain orchids one feels as one feels about human aristocrats, of whatever social rank, their affinity with the past, with a tradition surviving from an older world. Nor is this altogether fanciful, for the orchids are... an extremely ancient family, distinguished from other orders of plants not merely by their outward beauty or oddity but by the long and complex history of their floral evolution'.

But the Monkey Orchids aren't the only floral spectacle on the hillside. During the 1997 and 1998 seasons, the people monitoring the Monkeys discovered the leaf rosette of a Lady Orchid, identified in 2001 when it flowered for the first time.[2] Over subsequent years more appeared, but they looked different and people began to suspect there was something strange going on. It wasn't until 2006, when one flowered, that they discovered to their amazement that it was a hybrid of the Lady Orchid and the Monkey Orchid, the first of its kind known to have occurred naturally in the UK. Its official name is a tongue-twisting mouthful: *Orchis* x *angusticruris*.

Since then the hybrids have thrived and now produce well over a hundred flowering spikes each spring, a figure that seems to increase every year. They are remarkable plants, swarming across the slope and turning the hillside pink. Morphologically, they appear more or less intermediate between the two species, the inflorescences consisting of

---

[2] Molecular analysis by Bateman et al. (2008) demonstrated that Orchis purpurea at Hartslock is derived from Continental stock, arriving either on high-level air currents from the south, or more likely, by direct introduction by man (either deliberately or not).

heavily coloured, steroid-ridden Monkeys perched on the tall, robust stem of a Lady Orchid. They are enormous plants, full of teenage vigour.

A paper published by scientists at Kew Gardens in 2008 describes analyses that demonstrate that the hybrids are genetically closer to *purpurea* than *simia*. Their evidence suggests that the Lady is, pertinently, the mother, while the Monkey is the father.

While perusing the hybrids myself, I couldn't help but notice that there were far more here than there had been four years before. And that made me wonder. What if this hybrid population was expanding exponentially, slowly outcompeting its parents? The Lady Orchids would surely disappear first, but were there enough Monkeys on the hillside to cope? Even if the hybrids didn't outcompete their parents, would back-crossing[3] occur? This would pose a rather interesting problem for conservationists. Do you protect this incredibly rare hybrid, a plant that grows nowhere else in the country, or do you suppress it so that it doesn't wipe out (or genetically contaminate) the one remaining population of Monkey Orchids in the Thames Valley?

The Kew scientists once again offer an opinion. They've found that the hybrids do produce some fertile pollen and seeds, raising the possibility of back-crossing. Gene flow in this way would 'contaminate' the Monkey Orchid gene pool, blurring the lines between parent and hybrid. This has occurred in the past, as the study shows that the Monkeys have some DNA usually associated with Military Orchids,

---

[3] Back-crossing is the fertilisation of a parent (Monkey or Lady) with fertile pollen from a hybrid.

a third species, which had jumped across during a hybridisation event hundreds of years ago.

To top off this confusing scenario, the Lady Orchids at Hartslock have been shown to be more closely related to continental than Kentish plants, arriving either on high-level air currents from the south, or more likely, by direct introduction by humans – either deliberately or not. Some conservationists might therefore be inclined to order the swift removal of all the hybrids and Ladies from the reserve. However, the authors take a more optimistic view, as voiced by the reserve warden Chris Raper: 'It is my theory that in the past the three species [Lady, Monkey and Military] grew in colonies scattered all along the south Chilterns... They probably hybridised much more frequently and the resulting plants were consequently harder to split into three distinct species. Far from being a problem, the new hybrids might actually be returning the population to a more natural state where occasional mixing of genes between the species was normal.'

They look like they are here to stay, then. I personally agree with the view of the authors, as the hybrids are clearly successful and not currently doing any damage to the native Monkeys. The introduction of a few genes from such strong, healthy plants could help the Hartslock Monkey Orchids to rediscover their vigour. If anything, the effects of the sustained presence of both the hybrid and the Lady Orchids at Hartslock will be very interesting to watch over the coming years.

During the next hour or so a steady trickle of people came and went. Some were armed with enormous cameras on ostrich-legged tripods. Others would modestly produce a small compact to take a photo before hurriedly hiding it away again, as if they were embarrassed by the size of it. Watching orchid hunters is actually quite entertaining.

I sat and picnicked on the hillside, away from the Monkey Orchids and their admirers, with a wonderful view out over the surrounding Oxfordshire countryside. Every now and then the warm buzz of insects would be interrupted by the sound of a train sliding by in the distance. Chaffinches were trilling from the hedges and sheep bleated in the fields down in the valley. A barge chugged silently past on the river below, cutting a wide 'V' shape into the otherwise glass-like water.

I sat and read my book, *The Wild Places* by Robert Macfarlane, content with my morning's work and perfectly at peace with the surrounding landscape. Nature has its ways of communicating with the soul, and I was really benefitting from spending so much time outdoors, particularly after such a long winter. Eventually, after a great deal of internal persuasion, I dragged myself back down to the car, passing once again through the taped-off areas of orchids. I imagined tails curling around blades of grass as they swung through the sward, scampering mischievously up oats and bromes, each little monkey twisting and turning. I marvelled at every last plant, feeling sorry that I had to leave but assuring myself that I would return one day to see the Monkeys again.

Job Edward Lousley, known as Ted, was one of the best British botanists of the twentieth century and his contributions to the BSBI were invaluable. Lousley's interest in natural history began when, at the age of twelve, he was invited, along with a keen gang of similarly minded boys, to use the facilities at the South London Botanical Institute by W. R. Sherrin, then the curator. Sherrin would nurture their interests, taking them out on bicycles in the Surrey countryside to learn about natural history. By the time Ted was fifteen, he had already begun collecting specimens for a herbarium that eventually became the largest privately owned collection of its kind in the country.

In 1926, when he was nineteen, he was encouraged by Sherrin to join the Watson Botanical Exchange Club and then the Botanical Society and Exchange Club of the British Isles, which over time has evolved into the BSBI. His membership of such clubs allowed him to discuss his botanical interest with experts and the highly regarded collectors of the time. Before long, Lousley had developed an unrivalled knowledge of the British flora; he had slowly been adding to his herbarium and studying as much of the available literature as possible. He would also plan botanical expeditions accross the country, drawing up meticulous itineraries to ensure that he saw everything he wished to see.

At a relatively early age, his work on the genus *Rumex* (docks) and introduced plants earned him an international reputation and he made regular appearances on the radio. His wide range of publications include *Wild Flowers of Chalk and Limestone* in the Collins New Naturalist series and the *Flora of Surrey*. The latter documented his love for the county but remained unpublished until after his death in 1976. Lousley made a number of important botanical discoveries during his lifetime, the most noteworthy being that of the Military Orchid in Buckinghamshire. It was this population I had decided to try and track down.

The Military Orchid was first recorded in Britain at the

same time as the Monkey Orchid by Christopher Merrett in 1666. Early records (which are unreliable given the tendency to confuse *militaris* with *simia* and *purpurea*) suggest that the plant once grew in Surrey and Kent but, like the Monkey Orchid, its stronghold has always been in the Chilterns. It was once relatively common in this area, but the ploughing up of downland and over picking by collectors in the 1800s played a significant part in its subsequent decline. It steadily became rarer, its appearance mythical, until it died out altogether, probably when the last few plants were collected from a site in Hertfordshire.

The Military Orchid, sought after with such tenacity by Victorian collectors, passed into folklore. Orchid enthusiasts dreamed of its rediscovery. Years went by without even a murmur. Jocelyn Brooke, ever the optimist, wrote in 1950 that 'it is possible that *militaris* still lingers on in a few secluded spots in one or other of these districts' but even he, having heard no recent news of it, admits that 'it seems all too probable that this charming orchid has gone the way of scarlet and pipe-clay, Ouida's guardsmen, and all the other more romantic appurtenances of soldiering'.

This sad sentiment echoed the thoughts of many of *militaris*'s admirers, those who longed to see the deep roseine purple of this most royal of orchids once again in our woods. Little did they know, however, that it had indeed been rediscovered, somewhat by chance, at Homefield Wood in Buckinghamshire by Ted Lousley. He recounts the rediscovery in *Wild Flowers of Chalk and Limestone*, detailing the thirty-nine plants in the colony, eighteen of which had flowering spikes. 'In a way it was just luck,' he writes; 'the excursion was intended as a picnic, so I had left my usual apparatus at home and took only my notebook. But I selected our stopping places on the chalk with some care, and naturally wandered off to see what I could find. To my delight I stumbled on the orchid just coming into flower.' This day in May 1947 has become legendary, fast-tracking its way to fame as one of the most significant events in British orchid history.

Perhaps wisely, Lousley told no one the location of the newly discovered *militaris*, fearing that collectors would obtain this information and descend upon the site. It was of course searched for, but remained a secret for nine years, until it was eventually found by Francis Rose and Richard Fitter in 1956. They subsequently sent a postcard to Lousley with the enigmatic message: 'The soldiers are at home in their fields.' However, the secret remained closely guarded until 1975, when BBOWT, who had begun managing Homefield Wood in 1969, released the news that *Orchis militaris* had once again been found growing in Britain. A large photograph of the orchid appeared in the *Daily Mirror* on 11th June with the accompanying headline: 'The Beauty that Must Blossom in Secret'. It was not until the late 1980s that the exact location was released to the public.

Since its discovery in 1947, the colony has steadily increased in size, spreading from the Enclosure to the Meadow in 1983, and then to the 1985 Clearing[4] in 1995. The number of flowering plants grew to forty-five in 1995, and passed the one hundred mark for the first time in 2002. The Military Orchid has since been discovered at a second site in the Chilterns and at Mildenhall in Suffolk, where a

---

[4] These are the names given to the three distinct areas where the Military Orchids grow at Homefield Wood.

large colony grows in an old chalk pit. Work by Qamaruz-Zaman and colleagues in 2002 has shown that the three British colonies are genetically distinct and may therefore signify three separate colonisations from Europe. They are all incredibly important.

After the views and scorching slopes of Hartslock, it was nice to arrive at Homefield Wood and slip into the woodland rides, the air still warm from a day in the sun. On my arrival there had been five cars wedged into a small area near the entrance to the wood. It was obvious why they were there, of course, without even seeing their owners. Three of the cars were bumper 4x4s and that could only mean equally eye-catching Nikons and Canons.

Planes descended overhead on their way into Heathrow, but otherwise only the sharp cries of a couple of jays broke the silence of the wood. It did not take long to find the BBOWT nature reserve and then, shortly after, the path that took me into the Meadow. Despite the late hour, the heat from the sun remained strong and butterflies abounded, bustling from flower to flower, twayblade to trefoil. I was on my own.

Unlike the tiny Monkeys at Hartslock, the Military Orchid commands attention. I was immediately drawn to the stately rose and purple spikes towering above the other orchids. They were majestic, lordly, imperial; I couldn't come up with an adjective that would do them justice. Each enormous plant was an amalgamation of purples, holding a diadem of magenta flowers, a picture of pure power and strength. The resemblance to a soldier was striking, albeit one clothed in deep-pink and baggy pyjama bottoms; its pale, greyish-pink helmet appeared absurdly big but nevertheless gave the air of being prepared for war. I amused myself by inserting a short length of grass into the curl of its right arm: now it had a sword.

The first description of *militaris* comes from John Gerard's *Historie of Plants* in 1597: 'Souldier's Satyrion bringeth forth many broad large and ribbed leaves, spread upon the ground like unto those of the great Plantaine: among the which riseth up a fat stalke full of sap or juice, clothed or wrapped in the like leaves even to the tuft of flowers, whereupon doe grow little flowers resembling a little man, having a helmet upon his head, his hands, and legs cut off; white upon the inside, spotted with many purple spots, and the backe part of the flower of a deeper colour tending to redness. The rootes be greater stones than any of the kinds of Satyrions.' Amusingly, 'Souldier's Satyrion' literally means soldier's testicles. Of all the orchids known to Gerard, it seems *militaris* has the biggest balls.

I began noticing people walking down the track next to the nature reserve carrying the kind of camera equipment I would associate with the owners of Mercedes M-class and BMW X5s. They had to be coming from the Enclosure. Indeed, after five minutes of exploration, during which time I found three Fly Orchids by the side of the track, I entered another small paddock that had been fenced off from the surrounding woodland to discourage deer. Despite their grandeur, Military Orchids smell of cat pee, yet would still be a tasty snack for a passing doe.

I paused briefly in the Enclosure, putting myself in the shoes of Lousley. This was the very spot where he had made his serendipitous discovery sixty-six years before. I tried to imagine what he must have felt upon spotting a purple spike cast in a shaft of sunlight, just a short distance away from where his family and friends were eating their picnic.

I walked on, passing through the gate that took me into the final area, the 1985 Clearing, where the plants have flowered every year since 1995. It had an imperious presence here, and you could feel it. *Militaris* was lit up in the evening sunlight like paper lanterns. Its authority was tangible.

There were more Fly Orchids here, hundreds, and out in the open they were stunning, the light glinting off the band of iridescent blue-silver around the fly's middle. One plant had seven flowers and another four to come, making it the biggest I'd ever seen. Someone had begun to mark each plant with sticks but had clearly given up, so vast was the colony. It was encouraging to see this little orchid growing so abundantly, albeit under the protective gaze of its military guards.

Orchids, indeed many wildflowers, can form powerful associations with certain moments in our past. They will often grow in places with stories to tell, where you can feel a real sense of spirit and history. Jocelyn Brooke's love for orchids is epitomised by his obsession with the Military Orchid, which had completely and utterly captured his imagination: 'At this period – about 1916 – most little boys wanted to be soldiers, and I suppose I was no exception. The Military Orchid had taken on a kind of legendary quality, its image seemed fringed with the mysterious and exciting appurtenances of soldiering, its name was like a distant bugle call, thrilling and rather sad, a cor au fond du bois [a horn in the woods]. The idea of a soldier, I think, had come to represent for me a whole complex of virtues which I knew that I lacked, yet wanted to possess: I was timid, a coward at games, terrified of the aggressively masculine, totemistic life of the boys at school; yet I secretly desired, above all things, to be like other people. These ideas had somehow become incarnated in *Orchis militaris*.'

Sadly Brooke never found a British Military Orchid. He hunted endlessly, dreaming of finding this one plant and all that it represented for him. Had he found it, would he have realised that he already possessed the makings of a soldier? Time was to tell.

As impressed as I was with *militaris*, I must admit that the whole experience felt slightly fake. The Military Orchid encapsulates great rarity and beauty and as a result Homefield Wood has attracted orchid enthusiasts like moths to a flame ever since it became known as the site of the soldier. Unfortunately, good-natured as we are, we visitors constitute a significant threat to any place where rare plants have been sighted, trampling grass paths around fields and woods, and, full of botanical exuberance, often failing to spare a thought for the surrounding vegetation. Here in Homefield Wood, many individual orchids are now enclosed in open-topped chicken-wire cages, giving the somewhat bizarre impression of being on the local allotment.

I had, perhaps naïvely, always associated rare orchids with locations steeped in history that are truly wild. I'd come across such places in the Hampshire hangers and the Burren, but for the most part it is but an idyllic dream. Visiting Hartslock and Homefield Wood made me realise that as nice as it would be for wildness and rare orchids to go hand in hand, the constant stream of admirers during the flowering season could never allow this to occur.

The Desirable Category of Very Rare Orchids

Rare orchids are mollycoddled. We fence them in to keep the deer out, we graze their paddocks to optimise growing conditions; we even have to provide them with protection against our own elbows and knees. Sometimes they thrive in their barricaded sanctuaries, sometimes we get it wrong, but the unfortunate necessity of these practices can make their environments seem alienating and unnatural. Nevertheless, at Homefield Wood, conservation work has been an unrivalled success, and the Military Orchid marches on.

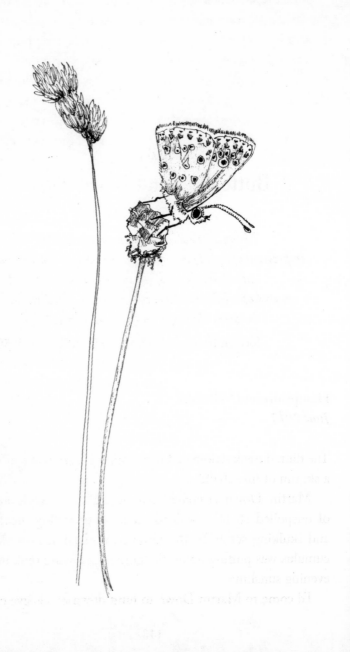

8

Butterflies and Burnt Tips

> *'Many of the British species give an odd*
> *impression of hardly belonging to the vegetable kingdom*
> *at all; like some of the lowest forms of*
> *microscopic life, they seem to exist on the borderland*
> *between the plant and animal worlds.'*

Jocelyn Brooke, *The Wild Orchids of Britain* (1950)

Hampshire and Wiltshire
June 2013

The flinted track slalomed down from the crest of the hill like a ski run of fine chalk.

Martin Down stretched out before me, a wide sweep of unspoiled chalk downland made up of rolling meadows and skulking scrub. In the distance, a marshmallow sky of cumulus was puffing up on the horizon, glowing pink in the evening sunshine.

I'd come to Martin Down to hunt down an elusive chalk

grassland specialist: the Burnt Orchid, *Neotinea ustulata*. Its little spikes of white flowers come wrapped up in wine-red buds, giving the tip of the plant a scorched look. If you find one you know you're in a special place as it's notorious for growing only in the oldest, most species-rich downland.

I'd tried and failed to find Burnt Orchids once before. The Friday after I'd returned to England from the Burren, my parents both had the day off so I'd driven us out to Clearbury Ring, just south of Salisbury. Overlooking the Avon valley, Clearbury is an Iron Age hillfort high up on the Wiltshire downs. Despite the altitude the morning was hot and clammy. My mother had pointed out that it was unheard of to have to wait until the very last day of May to get a proper May Day.

The day before had been wet and the clay soil of the arable field we'd walked along was soggy, squelching under our feet; the air was thick and humid. The hedges were sprinkled with snowy-white hawthorn and the rosy pink of dog-roses. They were lush and bushy, bulging with the new growth of spring. I was excited: this was the first orchid hunt of the year on my home turf.

The downland was booby-trapped with rabbit holes, each one guarded by a troop of purple thistles. We made our way through the grasses, eagerly anticipating little red-and-white spikes. But the further we went, the less likely it seemed we were going to find any; the grassland we were walking through was too species-poor. There were no iconic chalk grassland plants to be seen. Instead, the grass was filled with fluffy clovers, medicks and a clumpy grass called cock's-foot.

Calcareous grassland occurs on shallow, nutrient-poor soil on a bedrock of limestone, usually chalk. The soil is infused with calcium. It's famous not only for its wildlife but also for the White Horses carved into hillsides across the south of England. Chalk is a porous rock, so rainwater drains away very quickly and as a result the overlying grassland is usually very dry. It is a harsh environment for a single species to be able to dominate. As a result, it plays host to a multitude of different species. Typically found on valley slopes, it also develops in abandoned quarries, along roadside verges and railways generally across the south of England from Dorset to Kent and north into the Chilterns and Cotswolds.

One of the most biodiverse habitats in Europe, one square metre of chalk grassland can hold up to forty species of flowering plant. It's even been referred to as the temperate equivalent of tropical rainforest. Here in Britain, it supports a vast array of wildflowers, many of which can't be found anywhere else. This follows for insects too: Duke of Burgundy and Adonis blue butterflies, for example. Orchids, like many chalk specialists, have spent centuries adapting to this nutrient-poor environment. They survive in competition with equals.

Grasslands have been around for thousands of years. In Britain, they've been unintentionally encouraged by woodland clearance by humans and the grazing of wild animals. The practice of grazing livestock has shaped our wildflower-rich grasslands over the centuries, and its decline has played a significant role in the disappearance of much of our chalk grassland since the Second World War. Most wildlife present in lowland calcareous grassland can't tolerate agricultural intensification. As nutrient levels in the soil increase – often as a result of rain washing in fertilisers from neighbouring farmland – the orchids and their calcareous root-mates get bullied by rank grasses and clovers. What follows is a cascade of suffering: the insect populations, which are often completely dependent on some of the rare chalk plants, die out, followed by the birds and mammals which feed on them. Conservation bodies are working hard to halt the decline by bringing back traditional land management practices, grazing regimes and set-aside sites.

After breezing through the meadow, we arrived at a barbed-wire fence decorated with milky clouds of sheep's wool. On the other side was a steep south-facing slope home to a completely different community of plants. Instead of clovers there were vetches; instead of cock's-foot there was meadow oat-grass. The perfect habitat for orchids.

This type of grassland has a distinctive feel to it, an amalgamation of the senses: the trill of a skylark, floating down from some unseen coordinate in the sky; the aroma of thyme and basil, reminiscent of an Italian kitchen; and a pallet of colours flecked across a hillside, shifting softly in the breeze. I grew up botanising on the chalk, so this was a habitat I knew intimately.

I scaled the fence, managing to avoid getting caught in the barbed wire, and dropped down on the other side. My parents settled down in the grass and took out their lunch, making me promise to call them if I found anything. For the next hour, I walked a wiggly pattern across the hillside. Common milkwort twinkled blue, pink, mauve and white, taking shelter in the shadow of the earthwork ring. There were crowds of kidney vetch, red egg-shaped heads of salad burnet and the delicate golden rings of horseshoe vetch. The dainty white flowers of fairy flax filled pockets in the thatch and swayed drunkenly in the breeze.

I ducked down to look at the sugar-puff flowers of quaking grass, dangling like bait on a fishing line. Each floret was like a tiny squashed pine cone and every breath of wind made the whole panicle shiver. It grew everywhere on my way down to the bottom of the hill. Along with the skylarks singing overhead I could hear the scratchy call of a whitethroat from the scrub down below. A glance to my left brought a pair of mating dingy skipper butterflies atop a large head of salad burnet and, just behind it, a single Burnt Orchid! I leapt over an anthill and crouched low on the ground before disappointment sheared through me. What I'd mistaken for a Burnt Orchid was actually a white-flowered chalk milkwort.

I skirted a patch of scrub where blackthorn and hawthorn fought for a roothold. A brown butterfly plummeted from the sky and whizzed along the path before coming to a rest on a bright-purple Green-winged Orchid. It flicked its wings open to reveal a dusting of ginger: a Duke of Burgundy. I'd longed to see one of these little gems for years. As an insect-loving child who loved making lists, naturally I'd quickly caught the butterfly bug. There was a hillside near my house that I'd affectionately named Bentleigh Bank, after the farmhouse in the valley, that came alive with butterflies every summer. Each week I walked the same route around the downland, counting butterflies and submitting my data to Butterfly Conservation. Over the years, I'd clocked up more than thirty different butterfly species on the hillside, including brown hairstreaks, chalkhill blues and dark green fritillaries, but despite searching every spring, I'd never found the rare Duke of Burgundy. The male in front of me was a kaleidoscope of orange and nutty brown. I watched it for several minutes before it rushed upwards to chase off a peacock, a butterfly more than three times its size, and I lost it among the scrub.

My parents joined me and we continued to patrol up and down the paths, but there were no more orchids. 'You'd think they'd be easy to find, being rooted to the spot,' my father reflected. 'Surely you just turn up and there they are.' He was joking, of course. Having trailed after me on countless occasions while I looked for plants, he knew better than most how long it could take to find orchids. Or perhaps he thought I was just bad at looking.

After reaching the top of Martin Down, the track met Bokerley Dike, a long ditch that runs the length of the reserve and provides sheltered hollows for butterflies and, I suspected, Burnt Orchids.

I found my first Burnt Orchid here in 2010, hiding under a hawthorn near the car park. My father had been in Sweden visiting my grandfather, known to me and my sisters as Farfar, in hospital. Although I had GCSEs to think about, I'd persuaded my mother to drive me and my sisters to Martin Down for an evening hunting the Burnt Orchid. Esther had been the first to find one, and I remember being amazed at how tiny they were. Plants often seem huge from the photos in books. As we were admiring this little ruby orchid, we received the sad news that Farfar had passed away. The Burnt Orchid would always remind me of him.

I reached a crossroads in the track and began scanning the grassland, looking for the lone hawthorn tree that stood near to where I had seen the orchids three years previously. It took a few minutes but I eventually spotted it, smaller than I remembered, and started wading through the knee-high grass: this in itself wasn't promising. I searched for several minutes around the hawthorn but to no avail.

In 1886, A. D. Webster wrote that 'on some of the green sloping Kentish hills this little orchid is very abundant, and during the summer quite enlivens the landscape with its quaintly conspicuous flowers'. Today, the Burnt Orchid requires a lot of patience and prolonged searching; it certainly wouldn't be described as conspicuous. The last century has seen a dramatic decline in populations and range, thought to be primarily caused by the increase in ploughing after the Second World War and the decline in rabbit grazing following the onset of myxomatosis. Wiltshire, which has the Burnt Orchid as its county flower, is one of its few remaining strongholds.

Slightly disappointed, I decided to try another place, a half-mile walk across the downland. I crossed over the track and back into the grass, brushing through salad burnet bobs and the disc-like flowers of common rock-rose. A white-flowered group of milkworts appeared in my peripheral vision and I whipped round, heart racing, only to be disappointed again. I'd had enough of white milkworts.

I came across a middle-aged couple who were clearly botanising: they had hand lenses looped around their necks on threadbare ribbons and cameras slung over their shoulders. As I passed them, they were both crouched among the grass, the woman showing her husband some horseshoe vetch. It was a poignant moment; I wanted that.

After a few minutes, the grass shortened and I wandered into a large colony of Greater Butterfly Orchids, my unexpected twentieth. Greater Butterflies are ethereal plants: psychics, shrouded in mystery. Their tall, spindly stems arise from a pair of round, waxy leaves and hold an inflorescence of loosely arranged greenish-white flowers that glow eerily at twilight. Their Latin name, *Platanthera chlorantha*, is a tribute to this unique colour. Each flower has a long monkey's-tail of a spur that holds nectar for butterflies and night-flying moths. Unfortunately, there were only three or four plants in flower, the rest of them still in tight bud – it would have been quite a spectacle had they all been flowering. I guessed there must have been about 300 plants.

John Gerard thought very highly of the Greater Butterfly and was the first to coin its name. In his *Herball*, Gerard describes this species as 'that kinde which resembleth the white Butter-flie'. Despite deciding it had little or no use in medicine, he wrote that it had to be 'regarded for the pleasant and beautiful flowers, where with nature hath seemed to play and disport her selfe'. He found it regularly 'upon the declining of the hill at the North end of Hampstead heath' and 'in the wood belonging to a worshipful gentleman of Kent named Master Sedley of Southfleete'.

In *Flora Londinensis* (1777), William Curtis writes that 'the English name of Butterfly Orchis is scarcely warranted by the appearance of the flowers'. Seeing the butterfly in the flower certainly requires a significant stretch of the imagination. Instead, I thought, each flower reminded me more of an elephant. The white sepal on either side, presumably the wings of the butterfly, looked more like broad, flappy ears. The lip, supposed to be the butterfly's body, was the trunk. Finally, instead of tusks, two long diverging pollinia. The Greater Elephant Orchid doesn't have the same ring to it, though, nor the same connotations, so perhaps Butterfly is the right name for this handsomely delicate plant.

Known locally as the Night Violet in Wiltshire and as the White Angel in Somerset, the Greater Butterfly is a species best experienced after the sun has dipped below the horizon. Jocelyn Brooke, while considering which orchids he might stumble across in his local woodland, writes 'possibly the noble, lily-like Butterfly, which one should encounter preferably in the evening, when the air of the woodland is heavy with its fragrance. Perhaps there is no greater reward for the orchid-hunter than to come upon this beautiful flower at the end of his day's wandering'.

It once grew in the field behind our house. The Top Field, as we called it, was one of my favourite childhood haunts. I carried out endless butterfly transects and plant surveys and it was where my father and I went sweep-netting in the summer. At that stage, I'd only been botanising for a couple of years and was still finding my feet, getting to grips with the differences between the trickier species of common plants. I knew this was a butterfly orchid from the hours I'd spent reading my flower guides, but which one I wasn't sure. It was only returning to the orchid armed with camera and wild-flower guide that I identified it as Greater rather than Lesser. The pollinia were diverging, forming an inverted 'V', not parallel as they are in the Lesser Butterfly. This, among other subtle differences, appeared to be the key point of identification. Unfortunately, my delight at finding an orchid in the Top Field was short-lived. The following year, the field was ploughed up and converted into arable land and all that had grown there, Greater Butterfly Orchid included, was gone for ever.

I continued marching across the downs, clambering through an old earthwork that had left tell-tale lumps and mounds in the grass. As they have not been disturbed for hundreds of years, these are often the best places to find interesting plants. My search quickly proved fruitful as I came across the increasingly rare field fleawort, a plant that is related to the prolific ragworts so familiar to many of us as a weed of waste ground and gardens. Its flower is like a child's drawing of the sun, with yellow rays sparking from its golden centre. I found a Common Spotted Orchid just bursting into flower and large swathes of horseshoe vetch, adding yet another shade of yellow to the downland's palette.

I had subconsciously taken the white milkwort as a bad omen, likening it to the no-show of Burnt Orchids at Clearbury the week before. Unfortunately, it proved to be the case once again. I searched fruitlessly, failing to find the plants I knew were in flower. It was frustrating: somewhere, not far from where I stood, was my twenty-first orchid. I had experienced problems finding Dense-flowered Orchids in Ireland and now Burnt Orchids were troubling me too. I hadn't had much luck with these two *Neotineas*.

It was late so I headed back to the car. The sun was low on the horizon. Yellowhammers sang about a little-bit-of-bread-but-no-cheese from the hedgerow and, if I stood still and listened carefully, I could hear the soft purring of a turtle dove from the woods at the top of the hill. Despite failing to find Burnt Orchids for the second time in a week, I was in good spirits. I've grown up plant hunting on the chalk of Wiltshire and Hampshire and the experience of walking through chalk grassland on a warm summer's evening is something that I will always treasure. Seeing the Greater Butterfly Orchid had been an unexpected bonus. Jocelyn Brooke had clearly been in a similar situation himself: 'The Butterfly is not uncommon, and often provides a well-deserved compensation for the rarity which one has sought for, in vain, through the hot downland afternoon.'

The following afternoon, I drove through Salisbury once again, heading west towards Martin Down as I'd done the previous day. It was my last chance to see a Burnt Orchid before heading off to Yorkshire, so I'd decided to try a nature reserve near Coombe Bissett.

I knew before I started that this had to be a quick visit as it was my mother's fiftieth birthday. Several weeks before, she had issued strict instructions to keep her birthday free and ensure I could be around to help with the celebrations. I agreed to this without hesitating. No orchid hunting on the ninth. That was that.

By lucky coincidence, I wasn't up a mountain somewhere on the other side of the country when the time came. However, I'd also expected to have found the Burnt Orchid by now. So that afternoon, once all the guests had arrived at our house and I'd plied my mother with several glasses of champagne, I quietly slipped away. I would be back before she noticed anyway.

I turned off by the small church in Coombe Bissett and scaled the hill, almost missing the car park as I admired the lane-side display of doily-flowered Queen-Anne's-lace. I love this time of year when rural country lanes are lined with this white umbellifer.

Still dressed in my suit and tie, I grabbed my satchel from the back seat and locked the car. I headed off down the hill, surrounded by the dull green of arable fields. Coombe Bissett Down is a hidden valley that has escaped modern-day agriculture. Its slopes are too steep for machinery which, mercifully, has meant that it's been grazed annually by sheep during the winter, allowing the grassland to thrive.

Two years previously, I'd visited with Dominic Price to survey the hillside for Burnt Orchids as part of a BSBI threatened plants project. Dom is the director of the Species Recovery Trust, a small charity aiming to save fifty species from the brink of extinction by 2050. With a passion for Britain's plants and a shared sense of humour, we immediately became friends. Earlier in the day, I'd texted him to see if he was able to join me but, unfortunately, he couldn't make it.

The path, surrounded on both sides by a buttery wash of bulbous buttercups and common rock-roses, curved its way down to the valley bottom. I dilly-dallied along the path, marvelling at the wildflowers tripping over one another in their abundance. First came the small red-purple flowers of hound's-tongue, then horseshoe vetch in droves, irresistibly delicate quaking grass and bobbed heads of salad burnet. Burnt Orchids wouldn't want to grow anywhere else.

I paused, looking down at my ridiculous orchid-hunting outfit. Please don't let me be discovered here, I thought. With the ticking clock in mind, I hesitated no longer and plunged into the long grass. I made my way along the hillside, about a third of the way up, following one of the terracettes around the contour of the slope. These slumped almost-paths make steep pastures look rippled like a beach at low tide. It was the perfect place for Burnt Orchids.

Removing my jacket, I set about searching, enjoying the warmth of the afternoon sun on my shoulders. There were sheep in the adjacent field, outside the reserve. A lamb stood bawling in the middle of the field, and its mother responded from a corner by the woods.

I scanned the shorter patches of grass and spotted a white flower spike, the top few flowers scorched deep red. Here, at last, was a Burnt Orchid.

Burnt Orchids, or Burnt Tip Orchids as they are sometimes known, are for want of a much better word, cute. They are the teddy bear of the orchid world. Usually no more than ten centimetres tall, they are a wonderful sight among the grasses as the beautifully rich claret of the buds fades into white down the inflorescence. Some plants are tipped with a pale Burgundy; others with the deep colour of a Cabernet Sauvignon or a Pinot Noir.

Turning to my right, I let out a low whistle and walked over to an unusually large spike with a tall inflorescence holding at least forty flowers. Each lip resembled a dwarfed figure, as white as the chalk they rose from, dotted with burgeoning gunshot wounds.

I realised how easy it would be to miss, lost in the tangle of grass. It almost looked like red clover if you glanced over it quickly enough, or more whimsically, like little cherries. It's been described as a miniature edition of the Lady Orchid; however, while similarities exist, I felt the comparison hardly did justice to this exquisite plant.

The origin of the Burnt Orchid's Latin name, *Neotinea ustulata*, is debated. Some argue that it is named after a Sicilian botanist called Vicenzo Tineo, while others believe it refers to the supposed similarity to the African genus *Tinnea*, named by Dutch explorer Henrietta Tinne. Its species name *ustulata*, 'scorched-looking' or 'to burn', clearly refers to the dark colour of the unopened buds.

I wondered guiltily about the party back at home and quickly checked my phone. Just one text, from my father:

Where are you?

I grabbed my camera and stuffed it back in my bag. As I pulled my jacket on, I made my way down to the bottom of the slope, where I realised I was being watched by a woman and her dogs from the path. Stopping the dogs, she asked what I was doing and whether I was studying plants. Her whole face lit up when she heard I was looking for orchids.

'It's so nice to meet someone else who's looking for what you're looking for,' she said.

It turned out she lived in the village and enjoyed bringing the dogs here so that she could watch the orchids as they came out every year. I felt my phone buzz.

Leif, seriously, where are you? D.

She was giving me directions to a grassy bank within Salisbury Hospital that had sprouted sixty Bee Orchids the previous year. I was hurriedly noting down her instructions.

You better not be orchid hunting.

These Bee Orchids sounded huge! Some had had eight or nine flowers. I had to get the details down. After another couple of minutes, once I had made her confusing descriptions legible, I made my excuses and began jogging back to the car. Inexplicably, she hadn't asked me about my suit.

I arrived home and parked my car across the road. I ran up the drive and round the side of the house, hoping to avoid my father until I could make it obvious I'd been there the entire time. Straightening my jacket, I glided back into the party. I thought I'd seamlessly integrated myself until I bumped into my parents, who took one look at me and then almost in unison asked, how could I? I glanced down at my dishevelled tie, crumpled trousers and mud-scuffed shoes. Perhaps it would have been a good idea to change. It didn't matter though; I'd seen the Burnt Orchid.

9

The Lady's Slipper

'Though I know well enough
To hunt the Lady's Slipper now
Is playing blindman's-buff,
For it was June She put it on
And grey with mist the spider's lace
Swings in the autumn wind,
Yet through this hill-wood, high and low,
I peer in every place;
Seeking for what I cannot find
I do as I have often done
And shall do while I stay beneath the sun.'

Andrew Young, *Selected Poems* (1998)

Lancashire and Yorkshire
June 2013

Yorkshire, June 1930. Deep in the heart of an ancient oak wood, clothed in the fresh green of early summer, two brothers climbed steadily over knotted roots and crumbling scree. They walked in quiet tandem, ducking beneath branches that hung low with mosses and ferns. A long, arduous week weaving cotton in the factory had left them restless and tired. Because of this, they had almost stayed at home.

As they walked, bright shafts of light broke through in places, casting pools upon the woodland floor. The roar of the river, distant and monotonous, gave them their bearings. At the crest of a slope, the elder brother came to an abrupt halt. He pointed at a single flower growing on a grassy bank. Without meaning to, they had found Britain's Holy Grail: the Lady's Slipper Orchid was alive and well.

If ever the plant kingdom had a celebrity, it would be the Lady's Slipper. No other orchid has attracted so much attention and been subject to such levels of protection. Declared extinct in Britain in 1917, it was famously rediscovered in 1930 by the Jarman brothers, both cotton weavers from Silsden. This single plant remains the only one in the wild. Its stunning flowers are a feast for the senses: spirals of claret and bubbles of yellow. It is unlike anything else found in Britain. Almost mythical, it has become a household name among naturalists.

In 1629, John Parkinson, a London apothecary, found one 'in a wood called the Helkes in Lancashire neere the border of Yorkshire'. It was the first time the Lady's Slipper, or *Cypripedium calceolus*, was recorded in Britain. Over the following century, it was discovered across the north of England, growing in limestone pockets from Derbyshire to Cumbria, and Lancashire to Northumberland. If you wanted to see this plant in the 1800s, your best bet would have been to head to the Yorkshire Dales or Castle Eden Dene in County Durham.

Old records show that the Slipper was never a common plant in Britain, only rarely appearing in large numbers. William Curtis found it growing 'in considerable plenty in the neighbourhood of Kilnsey' near Grassington in the late 1700s, but there are few other records of large colonies.

An orchid of great beauty, with such large, conspicuous blooms, quickly caught people's attention; it was picked and dug up from the moment it was first recorded. In the late eighteenth and early nineteenth centuries, Victorian Orchidelirium was on the rise and the woods and hillsides were stripped of their Slippers at an alarming rate. They were transplanted to gardens, sold from market stalls in Skipton and Settle, and dried and pressed for personal herbaria. Today, there are more than 700 herbarium sheets of British *Cypripedium* specimens held in universities and institutions across the country. In 1976, Summerhayes lamented its decline in *Wild Orchids of Britain*: it could no longer be found growing at Parkinson's original location in Helks Wood, 'having been eradicated by a gardener at Ingleton who apparently had a ready sale for it'. By the mid-1800s, the Lady's Slipper was rare and in 1917 it was officially declared lost from Britain.

The Lady's Slipper fell into legend. Naturalists spoke of it in hushed tones, craving its mystical reappearance: would it ever be discovered again? Who would find it? And where? It became a plant of pilgrimage, an orchid that needed to be searched for and brought back to life. But Orchidelirium had driven the Lady's Slipper to extinction; or so people thought.

On 6th June 1930, quite by chance, a single specimen was discovered by the Jarman brothers on a remote hillside in Yorkshire, clinging precariously to existence. After much deliberation, they decided to share their find with a few trusted individuals, the identities of whom remain unknown. Upon realising that the Jarmans were telling the truth, the group retired to a local pub in the valley where they agreed to tell no one of its existence: they would protect it at all costs.

For decades, it was kept a closely guarded secret, known only to this small group of naturalists. The plant flowered sporadically, producing fewer stems, hoping one day to be pollinated by another plant. For nearly four decades, Willie Jarman and his son Bob kept detailed notes on the progress of the plant, taking measurements and counting flowers.

Many big-name botanists were excluded. Ted Lousley, for example, writes in *Wild Flowers of Chalk and Limestone* that the Slipper is 'one of the rarest and most elusive of British plants and one of the few I have not seen in the wild myself'. But a plant like this couldn't be kept quiet for ever.

In his book about the Yorkshire Dales, John Lee recalls an article by T. Hey published in *The Dalesman* entitled 'The Lost Slipper of the Dales'. Just after the end of the Second World War, Hey informs his readers that 'there have been occasional reports in more recent years that the Lady's Slipper still survives in these hills. Whether by some miracle an odd plant still remains it is impossible to say, for the botanists who may know wisely keep silent'. He goes on to recount an evening visit from Willie Jarman to discuss the local fox populations, but quickly realises that there's more on the agenda: 'I soon discovered that he had brought far more interesting news than that – he had chapter and verse for the existence of *Cyp-*

ripedium calceolus in the Dales. I will not disclose the spot, but he left me with notes of the history of those few plants over three years – how many had flowered, how cattle had nibbled back two of them, and how he was certain they still survived.'

In the 1960s, an article entitled 'Hunting the Lady's Slipper Orchid' appeared in *The Times*, detailing its discovery, though its source remained a mystery. In the following years, when the plant flowered, one member of the society removed the flower buds so that it would remain inconspicuous. Unsurprisingly, this proved unpopular with the rest of the society: any chance of the plant being pollinated and producing seeds had been squandered. One disgruntled member leaked the rough location of the orchid to the press. Mere days later, a hole appeared in the ground where the orchid had been. It had been stolen. Or at least that was their first thought. Purely by chance, the thief had departed with only part of the rhizome – an underground stem – while the rest remained present and intact.

The president of the BSBI, Edgar Milne-Redhead, called a meeting in the Dales in 1969 with representatives from the Nature Conservancy, the BSBI, the Yorkshire Naturalists' Union and the Yorkshire Wildlife Trust to discuss the conservation of *Cypripedium* and its habitat, as it was clear that protecting the remaining plant wasn't going to be a solution in the long term. They decided to pollinate the plant artificially, transferring pollen to the stigma using a pencil or paintbrush, but they would need to find plants of known British origin growing in gardens in order to cross-pollinate the wild one. The group met annually and became known as the Cypripedium Committee. Today, it has approximately ten members drawn from the Royal Botanic Gardens at

Kew, Natural England, the BSBI, the National Trust and the Alpine Garden Society.

Much earlier, in the 1950s, a friend of the Jarmans, who I will call Robert, had happened across a local plant nursery that was shutting down. In this nursery was a small bed with about twenty plant pots set in rows. It was early in the year and the pots appeared empty, so he casually asked the elderly owner what they were. The reply made him splutter: wild Lady's Slipper Orchids dug up from all over north Yorkshire. The collection had dwindled over the years, and these twenty pots were all that were left. Incredulous, Robert bought the lot and took them home. Over the years, he kept them growing in their pots, ultimately persuading some of them to flower. With the orchid's survival at heart, he eventually offered Kew some of his plants.

John Lovis, then a member of the Botany department at Leeds University, undertook the first attempt at pollinating the Lady's Slipper and in the first year, two flowers were produced. On returning to the site a couple of days after his first visit, however, he found that someone had snipped off both shoots. This was the final straw: the Cypripedium Committee weren't willing to risk the plant disappearing again.

To this day, every summer, from the moment the Lady's Slipper pushes its way up through the soil until the last of its seeds have been dispersed, the plant is under round-the-clock surveillance at a secret location. A warden lives in a small hut nearby and monitors the trip wires that surround the caged flowers. There are rumours he carries a shotgun. This level of protection is unprecedented for a plant, here or anywhere in the world. When inquisitive orchid lovers come asking questions, they are all turned away by the warden. The most desperate have been known to propose bribes: money, whisky and Wimbledon tickets.

No one is allowed near the plant. Even the members of the Cypripedium Committee limit themselves to a visit once every ten years. This isn't only to avoid attracting criminals: by getting close to the plant you disturb its habitat and begin compacting the soil, decreasing the chances of natural propagation occurring.

Other efforts have been made, besides protection, to save the flower. In the 1980s, scientists at Kew developed a skill that had been beyond the Victorian plant collectors: how to grow the Lady's Slipper from seed.

It took years of trial and error. A cultivated plant of British origin, affectionately known as the Hornby, was used to hand-pollinate the last remaining Lady's Slipper in the wild and is also now used to pollinate those being grown at Kew Gardens. Scientists at Kew, with help from a Swedish orchid enthusiast, developed a way to germinate the seeds without help from the Slipper's symbiotic fungal partner. It turned out that the Slipper was partial to pineapple juice. They found that seeds could be induced to germinate on an agar medium containing a very specific combination of vitamins and amino acids, watered down with some of Tropicana's finest.

Propagating orchids is notoriously difficult. And reintroducing the Lady's Slipper to our countryside from such a small sample of plants was always going to be particularly challenging. Today, the Cypripedium Committee, Natural England and Kew Gardens are working tirelessly on the reintroduction programme. In the first year, they prepared seventy-one seedlings to plant back out into the British countryside. Unfortunately, most of these either didn't survive or were found to contain genes from plants in continental Europe. The project was trying at all costs to preserve the indigenous British gene pool.

The Lady's Slipper is a long-lived species. The wild plant in the Yorkshire Dales is thought to be at least one hundred years old. A few have lived to almost two hundred. During development, everything must fall into place perfectly. Most importantly, the plant must establish a relationship with a specific fungus, to support it for the first few years of its life. It's not until the fourth year of the orchid's life that a green leaf emerges from the soil, which means that for the first three years of the plant's life, it is entirely dependent on the fungus. Only after a decade does the first flower appear.

To date, the reintroduction programme has been incredibly successful. The Cypripedium Committee drew up a list of suitable reintroduction sites: many were the orchid's former haunts across the north of England, while a few were new locations deemed to be appropriate to host this special plant. When the orchids were considered strong enough, they were planted out in secret and carefully monitored. In 2003, twelve sites had been established. In 2004, the first wild flowering occurred, at Gait Barrows Nature Reserve in Lancashire, and in 2008 the first naturally pollinated flower to produce a seed pod was observed.

Despite all the setbacks, the Lady's Slipper has clung to existence in Britain. Today, several of the reintroduction sites are open to the public so that every June, people can see this yellow-and-burgundy orchid flowering in the wild once again. Every flower that bursts into bloom is an enormous credit to the endeavours of the Cypripedium Committee,

Natural England, the scientists at Kew, cooperative land-owners and a long line of amateur naturalists who live and breathe orchid conservation.

High up on the hill, I watched as ravens floated by on thermals rising from the valley. Their croaks were barely audible over a cacophony of bleating lambs. Down below lay a tiny slumbering village; a string of greying houses marking a valley known to be a Slipper hotspot prior to the Victorian era. This seemed a good place to search.

There was something romantic about searching for the Lady's Slipper, hidden in the intimate rolling hills of the Dales, between quaint villages and tumbledown hamlets. Even today, very few people know the exact location of the last wild plant. So, unlike other rarities, it wouldn't be surrounded by white tape or people with cameras; and the surrounding vegetation would be lush and healthy, rather than trampled flat by botanists' boots. The Lady's Slipper lives a peaceful life, existing but not showboating, its flamboyant yellow flowers blooming in stark contrast to its humble, solitary nature.

I climbed further into the hills above the woods which covered the steep slope down into the valley. The upland grassland was awash with yellow buttercups and crosswort. Pignut bloomed white and lacey; sedges lined the paths with clusters of chocolatey-lime flowers, and wild thyme was beginning to add splashes of pink to the anthills.

The further I climbed, the more interesting the flora became. Unusual species began appearing out of nowhere.

As a botanist, I find the allure of plants is not just about the individuals, but the community as a whole and how it comes together to form a habitat. Here on the hillside, I had an exciting assemblage of species. There were purple and yellow mountain pansies scattered about in small groups, and the tiny star-like flowers of knotted pearlwort. Above me, on a steep bank where a spring had burst from the rock, were bubblegum-pink bird's-eye primroses. Early Purple Orchids, still in prime condition, were sprinkled here and there. Grizzled, age-old limestone broke free of the earth in places, with shallow grykes that housed shuttlecocks of male fern and hart's-tongue: shelter for the rock-hopping meadow pipits. I walked down towards the wood, listening to the eerie cries of the curlews on the moor.

Two days earlier, I had seen my first Lady's Slipper. I'd visited Gait Barrows Nature Reserve on the Lancashire–Cumbria border, where the first reintroduced plant had flowered in 2004. I'd arrived in the late afternoon, exhausted after a long, gruelling drive up the M6. For the final twenty minutes of my journey, I had been near-paralysed with fear as strange whining noises emanated from the engine of my car. But upon reaching the reserve, my worries vanished when I caught sight of a large photo of a Lady's Slipper tacked onto the gate. Below was emblazoned the message: 'follow the way markers to the viewing site'.

Gait Barrows is a wonderful mix of woodland and limestone pavement. As I walked to the viewing site, the early-evening sunshine was streaming through the surrounding vegetation. Glorious blackbird song flowed from the top of a birch tree. It had rained earlier in the day; the fresh smell of wet rock was heavy on the air.

I could see the limestone pavement through gaps in the trees, the same pale silver I'd enjoyed so much in Ireland. The pavements were different, though, lacking the wide, sweeping views of the Burren. Instead, you were never far from woodland. This created hidden pockets of pewter and green. The grykes were spilling over with bedstraws, speedwells and blue-green juniper. One was full of hart's-tongue fern, whose undersides were striped with sori: orange bars bulging from smooth fronds like the rungs of a ladder.

The path broadened, opening onto a wide clearing with small yew trees clustered in groups. Tufts of grass had fought their way through the limestone, giving the area an air of abandonment. A white tape was strung up in one corner, and below it was one of the most stunning orchids I'd ever seen.

It's difficult to describe the emotional impact. Over the years, I've read a lot about orchids and ogled hundreds of photos of their unmistakeable flowers, but nothing could have prepared me for that first glimpse of the fragile, jaw-dropping beauty of the Lady's Slipper.

A whole row of banana-yellow petals, unthinkably delicate, ballooned into shoes, slippers and clogs in front of me. Each one was hollow and smelled faintly citrusy. The base of the petal, closest to the centre of the flower, was shaped into a creamy tongue-like chute and speckled with small red flecks. Surrounding this absurd petal, in flawless complement, were four claret sepals that corkscrewed outward in a perfect compass.

I needed a moment to drink them in. I sat down next to a group of three plants whose flowers were lit up like Chinese lanterns in the evening sunshine. The warm light turned the slippers a golden yellow. The orchid closest to me had flowers with a wide sepal which fluttered in the wind like a main sail.

A few feet away was another cluster of Slippers hiding in the lee of a large limestone boulder, their flowers a burst of colour against the silvery-grey background. The day's rain was still evident: juicy droplets of water clung to the hollow petals as if frozen in place. I peered inside to see small pools collecting within. Not a good day for a swim if you were a bee.

These orchids are most frequently pollinated by small solitary bees of the genus *Andrena*. Attracted by the flower's scent, the bee lands on the lip and enters the slipper. After a minute or so looking around for nectar, it attempts to leave the way it came in, but finds the edges too smooth and slippery so is unable to escape. The only way out is via the two tunnels on either side, which are helpfully lined with stiff hairs for footholds. As the bee exits through the tunnel, it's forced to brush against one of the stamens, picking up pollen grains in the process. An elegant pollination mechanism for an elegant flower.

The graceful blooms have been inspiring folklore for centuries. According to one story, the 'Lady' was the Virgin Mary herself, and the red speckles inside the slipper are from the blood that spattered on her feet as she kept vigil beneath the cross. According to Medieval belief, the Virgin Mary would be protected by things of the plant world in her greatest hours of need. Hence her clothes were all made of flowers: hat, corset and shoelaces.

Another myth attributes the flower to the goddess Venus. Legend has it that while out hunting, Venus and Adonis got caught in a storm. Sheltering in a secluded spot, hidden behind bushes and undergrowth, they indulged in passionate love. A passer-by noticed Venus's golden shoe on the path, and when he reached down to pick it up, it was transformed into a slipper-shaped flower.

The name *Cypripedium* was coined by Linnaeus and derived from Cyprus, the supposed birthplace of Venus. The Latin epithet *calceolus* simply translates as 'little shoe' or 'little slipper'. All around me, delicate yellow moccasins bloomed in memorial of Venus. This quiet corner of Gait Barrows was her shoe closet.

Time seemed to stand still. Miraculously, I was alone; free to wander between orchids and marvel at each one. Their beauty overwhelmed me. I silently thanked the Cypripedium Committee for their remarkably successful efforts, and felt immensely privileged to be standing among Lady's Slippers growing in the wild. Ten years previously this simply wouldn't have been possible.

Buoyed by this sighting, I resolved to search for the one wild plant in the Yorkshire Dales. It might be futile, but I had to try.

Despite my best efforts, I had failed to learn the exact location of the last truly wild Lady's Slipper. Very few know where it grows, and those who do keep quiet. I had experienced the tight-lipped nature of its faithful guardians first hand. It is amazing, and slightly ridiculous, how well this secret has lasted the test of time. The Yorkshire Dales cover an enormous area and, while Summerhayes assured me that the Lady's Slipper once grew in these valleys, I had no idea where it still survived. Over the previous few days, I'd searched in woodlands and valleys across the Dales, but to no avail. It seemed the Slipper would elude me.

Tucking my thumbs into my rucksack, I gazed one last time down into the valley in this peaceful corner of the Dales, before plunging into the woodland. I squeezed through tiny gaps in the dry-stone wall, admiring water avens and listening to the willow warblers echoing around the canopy. Sun filtered down through oak and hazel. This was a wild wood. I quickly lost the footpath and found myself carving through a jungle of plants: solomon's-seal bloomed in shady fronds; bright-pink petals of herb-robert sprung from behind thick, swollen tree trunks; and pale snowballs of sanicle were dotted among the white enchanter's-nightshade.

Somewhere out there, hidden within the secluded folds of the Dales, the Lady's Slipper was waiting.

The wood was sombre and old. You could feel it in the gnarled crevasses of the bark and hear it in the birdsong muffled by moss and thick undergrowth. The rest of the world seemed so far away. This is what England must have been like thousands of years ago: a wonderful, unadulterated wilderness. I'd given up on finding the Lady's Slipper. Perhaps it was better if it remained undiscovered.

I was clambering along the hill, completely and utterly disorientated by the wood, with no idea which way I needed to go to get back to the car. After some time, twisting through the trees along the hillside, I still hadn't reached a footpath. I was very lost.

Eventually I reached an angled fence with a small makeshift gate. A battered wooden sign with the words 'NO ENTRY' daubed in red was slung over the barbed wire. I was struck by how out of place this was. I was convinced by now that the nearest public footpath must be miles away. It made no sense for a local landowner to have a gate here.

Leaning on the metal gate, I peered through the trees on the other side of the fence. The hazels and oaks were knotted and scarred, great stalwarts of the forest. Moss clothed the branches in furry green. Even at this stage, I never really thought I'd stumbled upon anything special, but after a couple of minutes of peeping and squinting, I did a double take: there, between the trees, was a small cabin painted dark myrtle-green, as camouflaged as a chameleon.

Heart beating faster now, I craned my neck, pacing up and down the fence. Could it be the wild site? Could I have just stumbled across a location so secret I'd found it impossible to track down through my network of sources? Keeping an eye on the cabin, I slithered carefully down through the ferns. Suddenly, I saw it: a flash of gold between two hazels.

It was the Lady's Slipper.

My stomach lurched. It was like nothing I had ever experienced. This was the very same plant found by the Jarmans in 1930, growing all on its own in the middle of this quiet wood. So many people had tried and failed to find this plant. I could only just see it, eleven bubbles of yellow visible between the trees in the distance. And there were the tiny corkscrews of maroon. The last Lady's Slipper. Inexplicably, I'd found its quiet sanctuary.

As I stood there, gawking at this impeccable plant, a movement beyond it caught my eye. A man was coming down the hill on the other side of the fence, walking slowly but purposefully towards me, zigzagging between trees: the warden. As he approached, every argument that I'd quickly prepared floated gently away, leaving my mind blank and my heart racing.

"'Ello,' he said in a thick Yorkshire accent, 'ah yer lost?' He seemed friendly enough and, I noted with relief, didn't appear to be carrying a shotgun. I stuttered, unsure what to tell him, so I gestured in the direction of the plant and mumbled something about orchids.

A smile spread across his face. 'Well now, well done for getting this far. Not many do.' My heart doubled in pace; was he going to let me see it up close? He was clearly under the impression that I'd known where the site was all along; understandably too, given how unlikely it was that I'd arrived here by accident. But seeing the hope in my eyes, he shook his head slowly.

I was busily trying to process what was happening: thirty minutes before I'd been wandering around the woods not really believing there was any chance of stumbling across the Slipper; now here I was, twenty metres away from it, chatting with the warden as if this was a normal place to bump into someone.

For the next half an hour, I tried my best to persuade him to let me photograph the orchid. I tried everything, desperation creeping into my voice. But no matter what I said, he stuck to his word and remained there, barring the way forward. 'I'd love to take you up the hill and introduce you, but I really can't – it's the worst part of the job.'

How is it that one plant can secure the loyalty of so many people? Why, each summer, does someone spend weeks living out here in the middle of this wood, all alone and miles from civilisation? I wondered how far these people would go, and what they would be willing to do, if confronted by someone trying to steal it. Such is the strength of the hold it has over them that even now, after all the success of the reintroduction programme, they remain unwilling to divulge its secret location.

And standing there, peering through the trees at that fragile, caged plant, I suddenly understood why. Each and every orchidophile involved in keeping the knowledge of the site secret had dedicated their lives to serving this plant. It had become a thing of genuine value, passed down over generations like a family heirloom. To reveal its location would be to welcome in the thieves and the vandals who care more for their own gain than the enjoyment of others. The idea of this delicate flower, nurtured lovingly for more than eighty years, at risk of being plundered after all it's been through, was a heart-breaking thought. That it must be so highly protected in the first place provoked a genuine sadness in me and, for that moment at least, I understood their reluctance to tell.

Defeated, I said goodbye to the warden and retreated, sneaking one last look at that mythical plant before I left. The warden went the other way, returning to his hut hidden deep beneath the trees.

Having experienced the wilds of the Dales for myself, I remain convinced there are more wild Lady's Slipper plants to be discovered in that part of north Yorkshire, an area little visited by anyone but the occasional sheep farmer or hardy hiker. The Dales are so wild and remote, and cover such a vast area, that it would be almost impossible to search it all properly.

Could there be more plants out there, just waiting to be discovered? In his 1948 work *Wild Flowers in Britain*, Robert Gathorne-Hardy tells a story of a Yorkshire mill-girl who went for a walk in the woods and returned with a bunch of forget-me-nots, among which was a single Lady's Slipper. When questioned, she couldn't remember where she had picked it. The exhilarating thought that a few wild Slippers are out there, growing undisturbed is captured perfectly by Summerhayes: 'What orchid enthusiast in Britain has not experienced the thrilling hope of re-discovering the rare Lady's Slipper in its erstwhile haunts, or even of finding it in a yet unrecorded locality.'

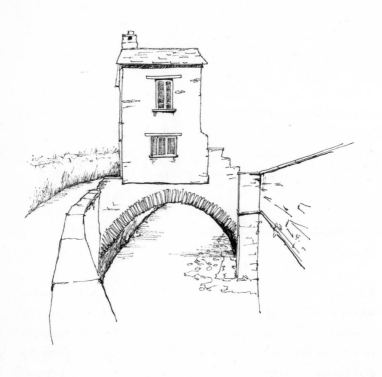

10

Curse of the Coralroot

'There is, about all orchids, something inherently
odd – something rather perverse and ambiguous,
something even a trifle sinister.'

Jocelyn Brooke, *The Wild Orchids of Britain* (1950)

Northumberland
June 2013

John Ray was an English naturalist of the seventeenth century
whose cataloguing of plants helped us take an important step
towards modern taxonomy. He was one of the first botanists
to comprehensively classify the British flora, travelling up
and down the country on horseback, discovering swathes of
wildflowers never recorded before.

Educated at Cambridge University, Ray developed a
longing to understand the natural world and swiftly became
fascinated by the wild plants growing in the countryside.
In 1656, he began working on what would become his

Cambridge Catalogue of plants, relying largely on what he had taught himself as there was a distinct lack of authority from which to draw. His work was the first attempt at classifying plants based entirely on science.

Once his *Catalogue* had been published, Ray hatched plans to put together a complete flora of Britain and would spend summers travelling across the country studying plants and their habitats. In 1659, he was accompanied by a friend on a tour of northern England, Scotland and the Isle of Man. Covering an epic distance in one summer, they recorded endless lists of new species. I'm envious of Ray in so many ways, but perhaps most because he had a botanical friend. In my decade of botanising, I had so far failed to find a friend who liked plants as much as I do.

An account of Ray's travels was published by the Royal Society. As well as detailing his findings, he included a travellogue for those who weren't particularly interested in nature. Ray was already trying to get more people interested in plants. Botanists have accumulated numerous stereotypes over the centuries: as crazy loners best left to themselves; as groups of fuzzy old scientists hiding away in herbaria filled with dried plants and microscopes; and members of societies whose average age is sixty-plus and who spend every day talking about miniscule plant structures in Latin. This is an image that must change. I respected Ray's attempts to make botany more accessible for those who had never shown any interest in plants.

Above all, Ray never took for granted the fact that he had been born into a time when the world was moving from being dependent on the teaching of Aristotle to making observations and testing theories with experimentation. From a botanical point of view, it must have been an extremely exciting time to be alive. In the twenty-first century, Britain has one of the most extensively recorded floras in the world so the chances of discovering new species are close to zero. Exploring the countryside in the 1600s, riding through woods bursting with plants, many of which have since been doomed to extinction, is every botanist's dream.

Fortunately for me, I didn't have to rely on horses to explore Britain. A journey that would have required a week for Ray to complete took me just a few hours by car.

After I'd finished with the Lady's Slipper, I'd driven to our small house in Haworth, a village made famous by the Brontë family. As my parents are both vicars, our Wiltshire home belongs to the Church, so we're allowed to live there as long as my father stays in his job. The house that my parents own is in Haworth, and has been the destination of our family holidays for as long as I can remember.

Slotted into the corner between two terraces, our tiny home is full of character: each room is different, and none of them are remotely matching. The kitchen has pine-panelled walls, sloping shelves and a random assortment of utensils hung on hooks. The floorboards are painted dark green, covered in the corner by a threadbare rug, and along one wall there's an old church pew. A rickety table holds generations of candles and vases of moorland grasses that have paled with age. Next door, there's a real hotchpotch of furniture in the mustard-yellow living room: a creaky wicker chair, a round wooden table once ridden with woodboring beetles and a lumpy sofa whose springs have all broken on one side. Above the breezy fireplace there's a collection of pine cones, washed-out photographs of my parents when they were younger and pencil pots filled with dried heather on the mantelpiece. Up the steep staircase is a strange, triangular room that I used to share with my sisters. My parents were across the hall in a room with gritty wallpaper and heavy, velvet curtains.

When I was a child, trips from Haworth cemented my obsession with nature. A walk along Bridgehouse Beck brought an up-close, personal encounter with a sleepy white-letter hairstreak, a butterfly that spends most of its life at the top of elm trees. One summer, my father and I spent a week birdwatching in boxy lakeside hides and saw avocets sweeping their curved bills through the water as marsh harriers circled overhead. On a trip to Fountain's Abbey the following spring, I saw my first Common Twayblades. Later, I would dedicate each holiday to botany, teaching myself how to identify the plants I wasn't able to find in the south. Spending so much time there played an important role in my formative years and the development of my interest in botany. This year the house in Haworth proved invaluable as a base throughout the summer as I yo-yoed my way up and down the country.

For the next two orchids on my list I had to travel further north, to Newcastle. They were the Coralroot Orchid and the Lesser Twayblade. These two species are small, inconspicuous and very good at hiding. I would be lucky to see them both in a day. Checking internet forums, it seemed like I was just in time. Both species had reached their peak and wouldn't be flowering for much longer. I had to see them in the next couple of days.

The evening before my trip to Newcastle, I was faced with an unexpected problem: my phone had switched itself off and was stubbornly refusing to start up again. I used the landline to call home and told my father, explaining my plan to drive to Northumberland and stay overnight at a campsite before returning to Haworth the following morning. It would give me two days to find both orchids. We agreed that it would be worth buying a cheap phone to see me through: a simple solution with no further worry.

If the broken phone hadn't been a bad omen, then the fact that Sainsbury's had sold out of cheap replacements should have been. I stood there in the aisle, weighing up my options. Spending seventy pounds on a phone for two days seemed excessive so I walked away, reassuring myself that I'd be back home before I knew it.

The drive up to Newcastle passed quickly and Fleur, my satnav, effortlessly navigated me through the city to Gosforth, where I pulled off the main road into a small layby outside Gosforth Park Nature Reserve. This small area of woodland in Newcastle is privately owned by the Natural History Society of Northumberland and is only open to members. I had managed to obtain a permit to visit the site for the day and it was here that I was hoping to see the Coralroot Orchid.

Corallorhiza trifida, to give it its Latin name, is so called because of the branched, coral-like appearance of the plant's 'roots' (technically these structures are horizontal underground stems rather than true roots). This bizarre orchid has no leaves, instead relying on its photosynthetic stem and fungal partner for carbohydrates. For much of its life it's almost completely dependent upon this relationship. Like the Bird's-nest Orchid, it's extremely fussy, and is only ever associated with fungi from the *Thelephora-Tomentella* group.

In turn, these fungi always form relationships with pines, birches, alders and willows. As a result, Coralroot Orchids are associated with two distinct habitats: damp, willow-carpeted dune slacks, and wet woodland filled with alder, birch and Scot's pine. The fungi obtain carbohydrates from the trees, only for it to be poached by the orchids. It's a one-way relationship: the parasitic Coralroot cheats the fungus, giving nothing in return.

I was cautious as I entered the reserve, very aware that without my phone I didn't have a copy of my permit. To anyone who didn't believe my story, I'd be trespassing. Or at least it would be what I like to call Good Natured Botanical Trespassing – when you see a rare plant on the other side of a barbed-wire fence and just have to get a look at it. In the past, I'd been chased away from rich wildflower meadows by landowners who hadn't understood my good-natured intentions. Hopefully I wouldn't run into a scary warden today.

A short way along the path into the nature reserve, situated just inside the treeline, was a squat wooden cabin. On the porch, two men dressed in khaki combat trousers and walking boots sat sprawled in camping chairs. Their faces were obscured by olive-green caps and one had a hand-rolled cigarette dangling limply from his mouth. Clouds of smoke hung in the air. Between them, on a tinny metal table, lay a pair of binoculars and a crumpled map.

They noticed me immediately and watched silently as I approached. It was intimidating. I greeted them and briefly outlined my predicament: no phone, no printer, no permit. I was trying hard not to think about how I would find the Coralroot Orchid if they turned me away. They still hadn't said anything. Eventually, the man with the cigarette stood up.

THE 52 SPECIES OF
BRITISH ORCHID

1.

1. Early Purple Orchid, *Orchis mascula*

Orchids numbered by their order of appearance in this book

2.

3.

4.

5.

2. Early Spider Orchid,
Ophrys sphegodes

3. Green-winged Orchid,
Anacamptis morio

4. Irish Marsh Orchid,
Dactylorhiza kerryensis

5. Loose-flowered Orchid,
Anacamptis laxiflora

6. Southern Marsh Orchid,
Dactylorhiza praetermissa

7. Common Spotted Orchid,
Dactylorhiza fuchsii

8. Man Orchid,
Orchis anthropophora

6.

7.

8.

9. Lady Orchid,
Orchis purpurea

10. Common Twayblade,
Neottia ovata

11. Early Marsh Orchid,
Dactylorhiza incarnata

12. Dense-flowered Orchid,
Neotinea maculata

13. Pugsley's Marsh Orchid,
Dactylorhiza traunsteinerioides

14. Fly Orchid,
Ophrys insectifera

15. Sword-leaved Helleborine,
Cephalanthera longifolia

9.

10.

11.

12.

13.

14.

15.

16.

17.

18.

19.

20.

16. Bird's-nest Orchid,
Neottia nidus-avis

17. White Helleborine,
Cephalanthera damasonium

18. Monkey Orchid,
Orchis simia

19. Military Orchid,
Orchis militaris

20. Greater Butterfly Orchid,
Platanthera chlorantha

21. Burnt Orchid,
Neotinea ustulata

22. Lady's Slipper,
Cypripedium calceolus

21.

22.

23. Coralroot Orchid,
Corallorhiza trifida

24. Lesser Twayblade,
Neottia cordata

25. Chalk Fragrant Orchid,
Gymnadenia conopsea

26. Lesser Butterfly Orchid,
Platanthera bifolia

27. Heath Spotted Orchid,
Dactylorhiza maculata

28. Heath Fragrant Orchid,
Gymnadenia borealis

29. Small White Orchid,
Pseudorchis albida

23.

24.

25.

26.

27.

28.

29.

30.

31.

32.

33.

34.

30. Pyramidal Orchid,
Anacamptis pyramidalis

31. Fen Orchid,
Liparis loeselii

32. Bee Orchid,
Ophrys apifera

33. Lizard Orchid,
Himantoglossum hircinum

34. Late Spider Orchid,
Ophrys fuciflora

35. Red Helleborine,
Cephalanthera rubra

36. Northern Marsh Orchid,
Dactylorhiza purpurella

35.

36.

37. Hebridean Marsh Orchid,
Dactylorhiza traunsteinerioides

38. Frog Orchid,
Dactylorhiza viridis

39. Musk Orchid,
Herminium monorchis

40. Dark Red Helleborine,
Epipactis atrorubens

41. Bog Orchid,
Hammarbya paludosa

42. Marsh Helleborine,
Epipactis palustris

43. Marsh Fragrant Orchid,
Gymnadenia densiflora

37.

38.

39.

40.

41.

42.

43.

44.

45.

46.

47.

44. Green-flowered Helleborine,
Epipactis phyllanthes

45. Dune Helleborine,
Epipactis dunensis

46. Lindisfarne Helleborine,
Epipactis sancta

47. Creeping Lady's-tresses,
Goodyera repens

48. Narrow-lipped Helleborine,
Epipactis leptochila

49. Broad-leaved Helleborine,
Epipactis helleborine

50. Irish Lady's-tresses,
Spiranthes romanzoffiana

48.

49.

50.

51.

52.

51. Violet Helleborine,
Epipactis purpurata

52. Autumn Lady's-tresses,
Spiranthes spiralis

'We believe you, mate, but we're going to have to take you to the orchids. There's no way you'd find them otherwise,' he said in a thick Geordie accent. A stroke of luck. He introduced himself as Paul, the reserve warden. Grinding his cigarette out on the table and readjusting his hat, he said goodbye to his friend and we began walking into the wood in uncomfortable silence.

My first impression was that it was gloomy. As we walked, the path became damp, then wet and eventually dipped into swampy pools. We stepped onto a boardwalk that took us through the waterlogged wood of alder and willow. It occurred to me that this was probably what the whole area once looked like: miles and miles of endless forest and swamps. Now this was all that was left and Newcastle was standing on the burial site of acres of wooded fens.

After walking for ten minutes, Paul turned off the path and made his way down to a birch tree in the reed bed where there were a few wooden boards laid out over the mud. I'd seen the Coralroot Orchids before he pointed them out, their bright-green stems glowing in the low woodland light. Miraculously, they were still in flower. After another minute of conversation, I thanked him for helping me and he left me alone in the middle of this dense, wet forest.

There were about thirty orchids altogether, spread out on cushions of moss among the reeds. The few dingy flowers were well spaced and pointing in different directions like a signpost. Each bloom had greenish-yellow sepals, held forwards as if they were offering each other an embrace, and a white lip that kinked downwards, spattered with crimson spots. While equipped for insect pollination, the flowers usually go about their business by themselves. The pollen is loosely attached and easily falls off, landing on the stigma and resulting in self-pollination. When fertilised, the flower is no longer needed and deteriorates. Because this happens routinely, the flowers are only at their best for a couple of days, making it a pesky species to catch in flower.

The first British record for the Coralroot Orchid was made by the Reverend John Lightfoot in 1777. He reported in *Flora Scotica* that 'Ophrys Corallorhiza' grew 'in a moist hanging wood near the head of Little Loch Broom on the western coast of Ross-shire'. At this time, 'Ophrys' was a general term used for orchids, not just the furry insect mimics that bear the name today. Since then it has been recorded across northern Britain, occurring as far south as Yorkshire, but mainly growing in eastern Scotland, where it has become the county flower of Fife.

Populations of Coralroot Orchid have been under-recorded and many have been overlooked completely. Their small, leafless stems blend into the background of moss and leaf litter. This, coupled with their erratic flowering behaviour, has resulted in many colonies growing undiscovered for years. Efforts to document its range have turned up a lot of new sites, predominantly across Scotland, where several large populations have been found. While it doesn't seem to be particularly vulnerable in its Scottish localities, drainage of its habitat and destruction of woodland could lead to a serious decline in numbers.

I'd never encountered a plant that looked as alien as these lime-green orchids. They popped up through the browned oak and birch leaves scattered across the ground, like six-eyed periscopes looking in all directions. Even now they were extracting nutrients leached from the birch tree they grew under like some weird foreign life form. Somewhere beneath where I stood was a large, intricate network of fungal hyphae that the orchid had somehow coerced into doing its dirty work.

Satisfied, I made my way back through the trees, along the boardwalk and back to the main road. Hopping back in the car, I gave Fleur instructions to direct me to a remote area of moorland south of Hexham where I was going to search for Lesser Twayblades. It was a forty-five-minute drive, meaning that if I found them quickly I might be able to return to Haworth before nightfall.

It was at four o'clock that afternoon, heading south on the motorway through Newcastle, that things started to go wrong. Noticing a small loss of power to the engine, I glanced down at the speedometer. The needle was slowly but steadily dropping towards zero. With a sinking heart, I turned off the radio and heard the last sound I wanted to hear: a sort of discordant grating. Without any warning, the car began juddering violently, scattering the pens and sweets piled up on the dashboard. I pulled into the left-hand lane and then into a small parking layby and ground to a halt.

I turned off the engine and sat still for some time, trying to think about what to do. The car had broken down. I was hours away from home. My phone wasn't working.

I felt incredibly foolish. My car had always been a disaster, so why had I attempted this journey without a working phone? I decided to let the engine cool down, hoping that it had just overheated. I read my book and tried to forget everything that had happened. To my right, the ceaseless roar of traffic flashed past. After half an hour, I tried the engine again, but was met by the same argumentative tone that I had heard before, as if it were saying: 'I'm old and tired, please just leave me alone'. I was completely on my own.

Time slipped by with the blur of motorway traffic. I suddenly came to, realising I'd fallen asleep. Deciding that there was no point just sitting here, and wishing I was at home, I slid over to the passenger seat and opened the door, letting in a blast of hot exhaust fumes. I looked up and down the road, searching for an emergency phone. Nothing. I packed a rucksack with food and water, my road atlas and a jumper, and set off back along the motorway.

The walk was awful. I was bombarded with hot dust and loud noise. Battered beer cans and crusted sweet wrappers collected at the side of the road and in the grass, thrown carelessly from car windows. Empty crisp packets scuttled and scudded with every rush of hot air.

After five minutes, I made it to a slip road and walked up to the roundabout at the top, noticing a walkway that ran overhead. I couldn't see either end. This was utterly ridiculous. Ducking under the bridge, I crossed no-man's-land and worked my way along the downward slope of the walkway until I could reach the railings. Shaking my head with disbelief at what I was having to do, I pulled myself up, climbing the metal railings and levering myself over the top and onto the walkway. Saluting the CCTV camera that stood watch above me, I turned and started walking into Newcastle.

One hour later, I was back in the car. After walking for twenty minutes, I'd managed to locate a pay phone. But I only had the coins for one phone call: the AA or home? In hindsight, the logical option would have been to call the AA, but I was too stressed and desperately wanted to talk to my family. There are no words to describe how relieved I was to hear my father answer the phone, his voice faint and crackly on the other end of the line. I knew it would all be okay. We agreed that while the engine could still start the best thing to do would be to try to get to the campsite I was booked into for the night. I would be able to sleep there and use someone's phone.

Now, back in the car, I felt more confident. The engine, while still not perfect, was sounding a lot better than it had before and the campsite was only half an hour's drive away. I could make it. Throwing the car into gear, I merged back onto the motorway. The needle of the speedometer was cruising towards seventy and for a moment I relaxed. The malfunction seemed to be over.

One mile later, I came off the motorway and stopped at the traffic lights at the top of the slip road. Able to focus again, I thought about how I was going to get from the campsite to see the Lesser Twayblades. It was too far to walk, but perhaps there would be a bus I could catch to the nearest village. Orchid hunting by public transport would be considerably harder.

Distracted, it was only as the lights turned green that I realised the Ford in front of me had its bonnet in the air. At least I hadn't broken down there. Checking my mirrors, I tried moving round it but stalled. I tried again but the same thing happened. Then the car wouldn't start at all.

I was suddenly aware of how dangerously problematic this was. I had broken down, this time seemingly for good, in the middle of a four-lane slip road at rush hour. Trying not to let panic take over, I flicked the hazard lights on and climbed out, carefully edging my way over to the car in front, as motorway traffic pulled up at the lights on either side. A frightened man in his forties sat behind the wheel, his view of the roundabout blocked by the grey wall of metal rising in front of him. He jumped when I knocked on the window.

'Hi, sorry, can I borrow your phone? I've just broken down too,' I said by way of explanation. He gave me a forlorn look before reluctantly handing over his phone. I didn't want to think about what I'd have done had he not been here.

I rang the AA, becoming increasingly frustrated with the automated voice on the other end. 'Before talking to a member of our team, we would like you to listen to the following safety rules.' I felt like shouting.

Suddenly, there was a roar behind me and I turned my head to see a large orange Range Rover pull up with the words 'TRAFFIC OFFICER' emblazoned across the bonnet. Pure relief flooded through me: here was someone who could sort me out. A uniformed man jumped out and strode over, looking cross. Hanging up on the AA, who were still asking whether I had used the correct fuel, I handed the phone back to the scared-looking man in the Ford. A pick-up truck had arrived to save him. The traffic officer took control of the situation. Without talking, he attached the Range Rover to the front of my car before ordering me back inside and towing me to relative safety on the verge of the roundabout.

Less than forty-eight hours after breaking down in New-castle, I was back. The events of the previous two days had been a sleep-deprived blur. The AA had towed me back to Winterslow in stages, travelling through the night. I had the car dropped off at the garage and walked across the fields back to my bed. I was in a bad mood: I had missed the Lesser Twayblade and without my car there was no way I would be able to travel anywhere remote to find it.

Two hours later, I was woken up for Sunday lunch by my father, who outlined his plan: he wanted to drive me up to Haworth that afternoon, and then on to Newcastle the next day. Lesser Twayblades were back on.

The Lesser Twayblade is predominantly a plant of Scot-land and northern England but it does also occur in Wales and in select locations in the Midlands and, quite randomly, on Exmoor. A much smaller plant than its relative the Common Twayblade, I'd decided to maximise my chances of finding this inconspicuous orchid by visiting a site in the north where it supposedly flowered in large numbers. This remote area of moorland didn't look particularly special.

I pulled on my wellies, slung my satchel over my shoul-der and began squelching down the hill, my father following close behind. I was very pleased to be able to share one of my trips with him and determined to make the long drive worth it. It had been a while since we had last been on a father-and-son trip. He was always a welcome addition, providing me with information about the birds and insects I might have otherwise ignored.

I described a Lesser Twayblade to him as best I could: petite, a basal pair of oval leaves, a reddish stem and tiny dark-red flowers. It was always harder searching for something you had never seen before. Lesser Twayblades are one of our most diminutive orchids. Not only are they small and difficult to spot but they tend to grow under the heather rather than out in the open, preferring the damp mossy areas to the dry peat.

The first British record for this miniature orchid was made by Christopher Merrett in 1666 from 'neer the Beacon on Pendle Hill in Lancashire'. He calls it *Bifolium minimum*, which efficiently describes its size and pair of leaves. The leaves actually grow halfway up the stem, but this is often difficult to observe as the bottom half of the stem is usually buried in the moss.

In the end, it took only a few minutes to find the orchids. I had crouched down to admire a scrambling web of wild cranberry when I spotted the first. The tiny flowers are forked like a snake's tongue, and a rich ruby red that contrasts beautifully with the bright green of the leaves. An alternative, and far nicer name is the Heart-leaved Twayblade, which is reflected in the species' Latin name, *Neottia cordata*; *cordata* meaning 'heart-shaped'. Together with the Common Twayblade, it used to be part of the genus *Listera*, which honours Martin Lister (1639–1712), the English doctor and naturalist, who was a friend of John Ray. However, genetic analysis suggested that it should be united with the Bird's-nest Orchid in the genus *Neottia*, meaning 'nest of fledglings', which is where it lies today.

Once we had got our eye in, there was no stopping us. The moor was overflowing with Lesser Twayblades. Amid exclamations of 'and another one!', we picked our way along a small stream that ran down the hill. Each cushion of *Sphagnum* moss hosted scores of twayblades. At one stage, I pulled back a sprig of heather to reveal eight, nine, ten plants. And another, and another. They were everywhere, spilling out from under the heather and bilberry bushes and gathering in the moss by the stream.

Lesser Twayblades are the court jesters of the orchid world. They seemed to congregate in groups and I imagined them telling stories, jokes and tales of mischief.

The smell of wet peat filled the air as I knelt down with my camera, water instantly soaking through to my knees and trickling down into my wellies. Every time I crouched in front of what I thought was the perfect specimen, it would move slightly, completely ruining the composition. It was as if they were purposefully trying to be difficult. I could imagine them teasing me, scampering back under the heather each time I got close enough to take a photo.

My father was relieved to have found them and we spent a happy couple of hours, working in tandem and pointing out the best spots to one another. I realised how much I missed the long summer days when he and I would abandon whatever we were doing to go and look for wildlife. There is a childlike freedom about giving yourself over to the excitement of nature, and being able to share that with someone else is doubly rewarding. Not for the first time, I thought about how fun it would be to share my travels with a friend my own age. Setting out to see all the orchids in Britain had proved to be a relatively lonely task. I was desperate for a friend who shared my love of plants.

The drumming of a snipe broke the silence. We spotted it whirring low over the bilberry: its didgeridoo-like call was utterly bizarre. The moor was beginning to bruise as concentrated patches of heather burst into flower. It wouldn't be long before the entire hillside was a sea of bright purple.

It was a relief to have the whole breakdown fiasco behind me. Two days previously, sitting in my Vauxhall on the roundabout in Newcastle, I hadn't expected to be on the moors searching for Lesser Twayblades so soon. During a particularly low moment, I had more or less resigned myself to the fact that I wouldn't be able to make it back up north in time to see them. My father really had saved my whole summer. I couldn't have done it without him.

At the time, I'd been too preoccupied with the Lesser Twayblade to think about the car, but now that I'd ticked off my twenty-fourth species, my thoughts turned to how I was going to see the next ones. How was I going to travel around? My parents certainly didn't have the time to ferry me everywhere. As the clouds gathered above, we bade goodbye to the jesters on the moor and made our way back to the car, beginning the long journey south.

11

Finding the Fragrants

'As Orchids are universally acknowledged to rank amongst
the most singular and most modified forms
in the vegetable kingdom, I have thought that the facts to
be given might lead some observers to look more
curiously into the habits of our several native species. An
examination of their many beautiful contrivances will
exalt the whole vegetable kingdom in
most persons' estimation.'

Charles Darwin, *The Various Contrivances by which*
Orchids are fertilised by Insects (1862)

Hampshire
June 2013

To say Francis Rose knew his plants is something of an understatement. In fact, he was one of the best field botanists of the twentieth century and he was constantly looking for excuses to go botanising. Writing in the *Guardian* in 2006,

David Bellamy remembers that whether it was 'flowering plants, ferns, mosses, liverworts and lichens – not a single species was missed by the mastermind. I can well remember, while high on Ben Lawers in Perthshire, Francis looking at his watch and exclaiming that, as dinner was already being served at the youth hostel, we could carry on for another couple of hours. We did, and wrote up our field notes before falling into bed, ready for an early start the next day. This was the infectious enthusiasm of the man'.

In 1981, he published his *Wild Flower Key*, an identification guide that took British botany to new levels, which is still in print today and used by botanists across the country.

Born in south London, Rose developed his interest at an early age, during long walks looking for plants with his grandfather. At school, he went about teaching himself biology as it wasn't offered on the curriculum. After getting a botany degree from the University of London, he eventually became a lecturer on the subject. He completed his PhD on lowland bogs in Britain, and it was while doing this work that he began to notice differences among populations of Fragrant Orchid. Perhaps there were more than two different types.

That first week after my trip to Northumberland was one of the slowest of the summer. My car was finished: it had broken down for the last time and was being sold for scrap metal. I'd grown attached to my clunky Vauxhall, and was sad to see it go. Emptying it of all my things was a low point.

I didn't have long to mourn, though, as a replacement was needed quickly. Reports of Lesser Butterfly and Heath Fragrant Orchids were coming in thick and fast. Fortunately, I was once again in the wonderfully capable hands of the Horner family who run the local garage. They have bailed me out on more occasions than I can remember and without them I would never have been able to spend the summer hunting orchids. They pointed me in the right direction and, after reluctantly emptying my bank account of all the money I'd saved from my job at Waitrose over the winter, I drove home in a silver Ford Focus.

In the following few days, I made a series of short day trips from home, enjoying the feel of the Focus on the road and slowly introducing it to some of my favourite botanical haunts. I coasted along empty country lanes, windows down, botany books flapping in the breeze, stopping only to check out road verges resplendent with wildflowers.

I hadn't intended to make another trip to see Burnt Orchids; I'd presumed they'd finished flowering by now. But the day I bought my Focus, I received a tip-off from a former primary school teacher of mine that the *ustulata* were still staging a wonderful display up at Martin Down. The accompanying photos sold it to me. This was the perfect excuse to try out my new car and make a second visit, with a better idea of where the plants were.

When I arrived, Martin Down was bathed in beautiful early-evening sunlight. Skylarks and whitethroats sung from their vantage points in the sky and the bush, enhanced rather than drowned out by the monotonous chirping of grasshoppers and crickets from the sward. I took my time, wandering slowly along the springy paths and stopping every now and then to take photographs of the hundreds of Common Spotted Orchids that grew in small pockets all over the downs. Some plants were small, weedy, with spikes bearing three or four mauve flowers, while others were much larger, their robust stems supporting upwards of fifty. Each flower was crowded with the hypnotic pink and purple swirls of raspberry ripple ice cream.

In my early days as a botanist, I collected plants. My mother, keen to encourage my interest, allowed me to pick flowers so long as they were common. I would take a few new species home after every walk, pressing them carefully between sheets of old newspaper and kitchen towel in her small flower press. After waiting several days, I would remove each flower and secure it in a notebook using sticky-back plastic, writing out key information underneath: name, date, location and habitat.

One June, I had been on a year-seven geography field trip with school. Like many children, I found pebble size and erosion rates exceedingly dull, so naturally spent the entire trip secretly botanising instead. When I found new species, I had to make sure none of my friends were looking before I picked them. With nowhere to keep the plants I collected, I had to eat my sandwiches two hours early in order to make room for fresh specimens of crosswort and field scabious. At the end of each day, I would have several new species stored in my sandwich box, but my prize find was a Common Spotted Orchid. Back at home, I excitedly took it out to show my mother. She was mortified. I had picked an orchid. I had clearly crossed some sort of line and was made to shamefully explain where it had come from. That was the one and only time I've ever picked an orchid. I still remember its purple swirls, now fastened lovingly into one of my notebooks.

The Common Spotted Orchids on Martin Down were fantastically variable. There were masses of them, springing up in groups of as many as nine or ten. They lined the path running alongside an old military earthwork that towered above the surrounding grassland, climbing up and over – utterly unfazed. Nearby, anthills were spilling wild thyme and fairy flax. Yellow rattle and horseshoe vetch glowed a warm yellow in the sunshine. As I walked, the summer's first brood of Adonis blues danced through the meadow.

Before long, I'd wandered into a small colony of Chalk Fragrant Orchids, interspersed among the Common Spotteds. Their flowers are babyish: soft, muted pinks and exaggerated curves. Some inflorescences were cylindrical barrels, others more like cathedral spires in the way they towered over the surrounding vegetation. I'd never seen Fragrant Orchids so large and dominant. Regardless of height, each delicate flower spike stood gracefully in the short turf, waiting patiently for the sun to dip below the horizon.

Fragrant Orchids have a very long, nectar-holding spur at the back of the flower and can therefore only be pollinated by insects with an equally long proboscis. Night-flying moths are ideally adapted for this and so it is at night, rather than by day, that most pollination takes place. As the evening draws to a close, when most other plants and their pollinators are turning in for the night, the Chalk Fragrants turn on the style. Within minutes of the sun dropping below the horizon, the orchids release an overpowering fragrance into the warm evening air that moths find irresistible. You haven't met the Chalk Fragrant until you've experienced it at twilight. Its afternoon aroma attracts butterflies, day-flying moths and the occasional botanist, but the orchid is merely biding its time and gearing up for the evening show.

Already slightly light-headed with the sweet, sickly scent of these orchids, I carried on, working my way down Bokerley Dyke. Then came chalk downland at its very best. I weaved through the grassland, sidestepping the little yellow flowers of field fleawort, the rare relative of the ragwort that grows in profusion later in the summer. In doing so, I almost trampled some dropwort, whose pure-white flowers were still tucked away inside swollen red buds.

About ten metres away stood a Burnt Orchid, tiny and well past its best but a Martin Down Burnt Orchid nonetheless. It turned out to be merely a sentry, a lookout, for on the slope below were a further fifteen spikes. Their wine-coloured sepals faded to pale strawberry at their base.

I was suddenly and unexpectedly overcome by the beauty of these richly coloured plants and the never-ending floral downland surrounding them; by the way the quaking grass danced in that ethereal evening light; and by how vulnerable each component of this unique community appeared.

There is something remarkably humbling about sitting in chalk grassland at sunset. I felt honoured to witness the movements of the down as everything silently and simultaneously slowed in the dwindling light. Butterflies finished their nectar meals and dropped one by one into the long grass; the thick, gloopy warble of a nightingale faded into the gloam; and the bumblebees, which had been busily buzzing, nestled into the chalk milkwort right in front of me.

Deciding to go and check on the progress of the Greater Butterfly Orchids that I'd found a fortnight earlier, I walked up the hill, parting a sea of upright brome as I reached the top. The colony was in fine flower, grand spikes of exquisite ghostly-white flowers open from top to bottom, glowing in the dusk. There were hundreds of them in the grass in front of me. Just like the Fragrants, Butterfly Orchids are predominantly pollinated by night-flying moths of the Noctuid family. As if on cue, the dark outline of a moth buzzed in, hovering before one of the flowers before swooping away and crash-landing in a grassy tussock, its wings vibrating so rapidly that they were reduced to a blur.

By now the sun had dropped and a greyish light had set over the downs. The scent on the air was pungent, overpowering almost. I sat in the middle of the colony, surrounded by glow-in-the-dark Butterfly Orchids, becoming giddier by the minute and totally, utterly happy.

Having spent the weekend with my family, Monday morning dawned, spelling the beginning of a busy week. I set out with my friend Ben to drive to the New Forest and hunt down some of the commoner orchids of Heath and Bog. The air had a midsummer warmth to it, the sky was a bright white blanket and Ben was on top form, keeping me entertained with tales of his first year at university.

As I had grown up in Wiltshire, summer trips to the New Forest were a monthly event. Over the years, I had fallen in love with the forest, with its wide-open plains dotted with gorse and small birch copses. I have fond memories of the wonderfully named Upper Crockford Bottom near Beaulieu where, early in my courtship of nature, I would go dragonflying with Dad, chasing hawkers and demoiselles up and down the tiny stream. During later visits, my attention was robbed by the indigenous flora and I would comb the stream for pillwort and Hampshire-purslane.

By the time I hit GCSEs, I'd already sought out a whole host of rarities: bastard balm at Wilverley, pennyroyal and small fleabane near Cadnam, and glorious manta-ray-blue marsh gentians south of Lyndhurst. During my A levels, I moved on to the trees and spent hours over the winter teaching myself how to identify them without their leaves. It's become a special place for me and provides a nice alternative to my beloved chalk downs.

I picked Ben up at ten and drove down to Wilverley. Most of my friends had left Salisbury, or 'The Shire' as it's colloquially known to departing students, to go to university. I'd seen very little of them during the past six months and was feeling left behind. They had moved on, made new friends and discovered a whole new sense of independence in cities across the country. I had a sense of being in limbo that the freedom of having a car could do nothing to assuage.

We pulled into the car park at Wilverley and looked out over the plain. There are few places in the New Forest that are as vast and flat as this; the grass ran uninterrupted into the distance, nibbled to within an inch of its life by wandering ponies. In late summer, field gentians and Autumn Lady's-tresses can be found in their thousands here.

The sun was beating down as we wandered across the plain, carefully avoiding the young foals and their protective mothers; it was yet another lovely day and with any luck I would be three more species down by dinner.

A happy-go-lucky person with a dry sense of humour, Ben was one of the first friends I made at my secondary school. We learned to play the clarinet together, though to the frustration of our teacher spent more time making each other laugh than playing music. As well as being an accomplished musician, he loves everything technical, so was always the friend to go to when your bike or laptop wasn't working.

While less enthused by nature, Ben shares my love of the outdoors and will go to great lengths to make sure he camps in the most far-flung places. Despite this, he remains terrified of horses, and kept flinching as the ponies moved closer to investigate.

We reached the other side of the plain and began walking through the heather, the parched ochre of last year's bracken crunching noisily beneath our feet. The smack of a stonechat drew my gaze to the gorse and, sure enough, there was a handsome black-and-red male sitting on top. We continued along a grassy path and soon came across the grey-green foliage of cross-leaved heath, showing the soil here was slightly damper, and then walked right into the colony of Lesser Butterfly Orchids we had come to see.

Against the dark background of the heather, these orchids were bright and greenish white, much like their greater cousins. A pair of waxy leaves gave rise to a spike of their signature weird, ghostly-pale flowers that are somehow both graceful and elephant-like.

The English epithet 'lesser' is something of a misnomer as, depending on the habitat, these plants can become taller than their 'greater' cousins. When growing in chalk grassland, the spikes appear short and compact, whereas in deciduous woodland they will be taller, with delicate, dainty inflorescences. Here on the damp heathland of the forest, the orchid's preferred habitat, they were somewhere in between. Interestingly, genetic analysis has shown that the two butterfly orchids are extremely closely related, suggesting that they have only recently split to form two different species. But while DNA sequence, height and form may not be the easiest characteristics for identification, the two can be easily differentiated by taking a look at the pollinia, visible as greenish lines within the flower. In the Lesser Butterfly, the two pollinia lie vertically parallel to one another, whereas the pollinia in the Greater Butterfly diverge from the top downwards, forming an inverted 'V' or 'U'.

Over a month into the season, I had retuned my eye and if I knew what I was looking for – and it was relatively common – I could generally find it from a fair distance away – something that comes naturally if you are doing the same thing over and over again but seems impressive from an outsider's point of view. Ben could not believe that I had seen the first orchid from so far away. This was also a defining moment: I had now seen twenty-six species, I was halfway there.

Aware that not everyone can spend hours looking at plants, I suggested we continue our tour of the forest with a visit to nearby Sway. Here, the heath bordered on wet boggy flushes that played host to the frilly-flowered bogbean and buttercup blooms of lesser spearwort, as well as hundreds of Heath Spotted Orchids, another new species for the year, looking extremely fresh and remarkable in their variation. I photographed tall plants with plain white flowers, tiny black-currant-coloured ones and many whose petals were daubed with calligraphic purple flourishes. Most were broad landing pads decorated with the dot-dot-dash of floral Morse code.

There are some extraordinary local names for the Spotted Orchids; in Wiltshire, they are variously known as Old Woman's Pincushions, Kite Pans and Dandy Goslings, while in County Durham, it is Scab Gowks, and in Kent, Skeat-legs. On the Shetland Islands, they are called Curlie Daddies and in Somerset, my personal favourite, one would refer to them as Choogy Pigs.

Seed production in the Heath Spotted Orchid perfectly exemplifies the outstanding survival methods of the Orchidaceae. Webster, writing in 1898, describes how one seed capsule contains approximately 6200 seeds; as a plant will often have at least thirty capsules, each individual can therefore produce a staggering total of 186,000 seeds. These large numbers may not mean very much, so to help visualise them we can use an example from Darwin's book on the fertilisation of orchids. Consider a single Heath Spotted Orchid. Allowing for 12,000 bad seeds, one individual would be able to produce 174,000 offspring. If each one was given fifteen square centimetres in which to grow, the progeny of a single plant would cover an acre of land – about the size of your average football pitch. If each of these 174,000 plants were to reproduce at the same rate, the resulting grandchildren would cover an area larger than Anglesey and the great-grandchildren of that one original plant would almost cover the entire landmass of our planet. So in just three generations, a single orchid has the potential to produce enough offspring to cover nearly all the land in the world. I looked around at the world-dominating assortment of Heath Spotted Orchids and wondered how many offspring they would be able to produce. Thankfully, competition from other plants keeps them at bay.

Once I'd wiggled into wellies, I began sploshing as carefully as possible across the bog, the ground sucking and gurgling underneath me with every step. My efforts were rewarded on the other side with the carnivorous pale butterwort, more Lesser Butterflies and then my third new species of the day: a Heath Fragrant Orchid. It was a weedy plant with four or five flowers perched on a small hummock that rose from the dark waters. Like the Chalk Fragrants I had found on Martin Down, its flowers were curvy and candy-floss pink. I knelt to sniff it: it was quite pleasant; spicy almost, like cloves.

This little orchid has only recently been recognised as a separate species. Originally classified within the composite 'Fragrant Orchid', it was first described as a distinct variety by G. C. Druce in 1918, before achieving subspecies status in 1991. Francis Rose was the key driver in unscrambling the Heath Fragrant from the Chalk and Marsh Fragrants, and I had been told by Steve Povey that it was here at Sway that he had made some of his key observations.

Rose gave them subspecies status in the first edition of his *Wild Flower Key* in 1981, but strongly implied they were worthy of specific status. Their full species status was finally confirmed by scientists at Kew Gardens, after DNA analysis showed strong support for the three species of *Gymnadenia*: *conopsea* (Chalk), *densiflora* (Marsh) and *borealis* (Heath), all named after their respective habitats. Unlike the other two, the Heath Fragrant doesn't require calcareous soils and so is most often found in upland grasslands in the north and west or in boggy areas in southern Britain.

Welly-less Ben, who had been playing on his phone while I was taking photos, seemed to be regretting his request to accompany me on a trip to see orchids. Many people can look at a plant for ten seconds, some for five minutes and very few for several hours. But having found all three orchids so easily, I could tell his mind was wandering: we had seen them and now it was time to go.

Although we'd known each other for eight years, this was the first time Ben had seen me botanising, and I could tell he'd found the whole experience quite bizarre. In *The Military*

Orchid, Jocelyn Brooke admits that he isn't a true botanist and once accompanied a 'real botanist to the Sandwich golf links, celebrated for a number of rare plants; it was a chilly afternoon in spring, and no weather to dawdle unless for a very good reason. My friend was in pursuit of a rare chickweed and every few yards would throw himself flat on his face and remain there, making minute comparisons, while the glacial sea wind penetrated my clothes and reduced me to a state of frozen irritability'. This experience makes him appreciate 'how boring botanists must be to non-botanists'.

I suspect Francis Rose, in his eagerness and infectious enthusiasm, rarely failed to get even the most indifferent companions interested. Though I never met him, his ceaseless desire to find the plants he was after, and his willingness to share the experiences he'd had with others, is truly inspiring. Rose contributed so much to what we know about the British flora, extending far beyond our native orchids.

12

Those Little Green Ones

*'For when a man falls in love with orchids, he'll
do anything to possess the one he wants. It's
like chasing a green-eyed woman or taking
cocaine... it's a sort of madness.'*

Norman MacDonald, *The Orchid Hunters:
Jungle Adventures* (1942)

Powys and Bridgend
June 2013

The successful outing to the New Forest with Ben had
brought me up to twenty-eight species for the year, but in
order to reach thirty, I would have to travel a little further
afield. The following few days were spent at home, juggling
family time with the careful planning of my upcoming trip
to Wales. It was exciting; the weather was holding out and I
was feeling confident.

I was at a friend's birthday on Saturday night and set

out alone the following morning, feeling slightly worse for wear. The M4 was heavy with the sluggish traffic heading for Cardiff so I crossed over the Severn Bridge and into Wales at a crawl. It was a relief to escape the white noise and exhaust fumes of the motorway and climb steadily up into the Brecon Beacons National Park. I spent the night in a small, family-run campsite lying in the shadow of the mountains, waking the next day to a bright, crisp morning.

I had come to Wales for two orchids I had never seen before, the Small White and the Fen, which are two of the most notoriously elusive plants in the country. I was going to start by visiting a small nature reserve in the heart of Wales that I had recently been told was a reliable spot for the Small White Orchid. I arrived late morning, quite literally in the middle of nowhere. The narrow road I had been driving along had gradually tapered into a single-file lane that meandered into the hills, taking me further and further from any form of civilisation.

To get to the reserve, I followed the footpath up a long track, before passing through the garden of a chalk-white farmhouse, opening and closing numerous gates along the way. To many urbanites, walking along a path through someone's garden might feel like an invasion of their privacy, and they tiptoe past as if walking through their front room uninvited. But in rural Britain, footpaths regularly pass through people's farms and gardens, a relic of past communities and their easy-going attitude to strangers and private land. A tiny stream trickled past at the side of the track, filled with sedges and water-crowfoots in cushions of springy mosses. The entrance to the reserve was unsigned, one more gate in the long line of opening and closing, latching and unlatching.

Those Little Green Ones

Ten metres into the damp, reedy meadow and the beauty of the place became obvious. All around me brilliant orange butterflies flitted, gliding along one minute, disappearing the next. They never stopped, or at least never long enough for me to get anywhere near them with my camera: small pearl-bordered fritillaries, one of the most striking and intricately patterned butterflies in Britain, and only the second time I had seen them. Every time I thought they would land, they would change their mind or meet another of their kind, engaging in a ferocious aerial battle that took them far out of my reach.

The meadow was surrounded by the steeply sloped hills of Powys and valleys that stretched as far as the eye could see before twisting away, to follow the flow of the river. Overhead, military jets were practising, tearing across the sky, leaving only a trail of noise in their wake.

The Small White Orchid, or *Pseudorchis albida*, is difficult to find simply because it never grows in large numbers. I had been lucky on my trip so far to have seen some rare plants but the swathes of Sword-leaved Helleborines in Hampshire and the army of Man Orchids at Darland Banks had made things relatively easy. Showy orchids are not hard to spot when they grow in such large numbers, no matter how rare they might be. The Small White, as its very name implies, is not particularly eye-catching and refrains from the egotistical displays seen in some of our other species. The handful of sites and accompanying directions that I had managed to acquire by no means filled me with confidence, ranging from having 'only one plant in flower', to 'three plants that have flowered for the last two years' and 'a couple of plants between two rocky areas in a west-facing basic flush'. These sites were some of the best in the country, but I knew it would be a tricky species to tick off.

I had reached the top of the nature reserve and stood admiring a sea of Heath Spotted Orchids; never had I seen so many in one place. This was a wildflower heaven. There were sedges everywhere, particularly in the wetter areas where the grassy hillside bordered on the reeds: pale, pill, star, carnation and oval. Wood bitter-vetch grew abundantly, dyer's-green-weed was just bursting into golden flower and Heath Fragrant Orchids, larger and far more satisfying than those in the New Forest, were dotted across the slope. Yet even more amazing were the bluebells; great purple carpets of them under the oaks at the top of the hill, spilling out into the grass and spreading down through the meadow.

As I watched, a shadow drifted across the hillside and looking up, I saw the large silhouette of a red kite as it rose above the treeline, banking to one side as if to show off its distinctive profile. Higher up still, a hobby raced over, a silent mimic of the jets now long gone.

While taking all of this in, I had noticed four off-white poles sticking out of the ground in the middle of the field. On approaching, I realised that each one marked a single Small White Orchid; tiny, dainty and unassuming. They were even smaller than I had imagined, consisting of creamy-white flowers rather loosely arranged atop a spindly green stem. Unmistakable. *Pseudorchis albida*, derived from the Greek for 'white-flowered false Orchis', presumably acquired its name from its superficial resemblance to the genus *Orchis*, but while appearing similar, it differs in many characteristics, particularly in the shape of its underground tubers (the checking of which is not recommended, because it is illegal to dig them up). The species has undergone many nomenclatural changes over the past 350 years and has been moved from genus to genus by seemingly indecisive botanists. *Gymnadenia*, *Habenaria*, *Orchis* and even *Platanthera* have all played host to this diminutive orchid.

The first British record for *Pseudorchis albida* comes from John Ray who, in 1670, found 'Orchis pusilla odorata radice palmata... on the back of Snowdon-hill by the way leading from Llanberis to Carnarvon'. In Britain, it is very much an upland plant, the majority fussily growing in undisturbed meadows on the well-drained, nutrient-poor soils of north-west Scotland. Many populations waned and became extinct during the twentieth century, resulting in the loss of Small White Orchid from 66 per cent of its historical range in Britain. This, coupled with its specific habitat requirements and small population sizes, has made it a deceptively difficult species to track down.

So that was it. I had found the Small White, supposedly one of the trickiest species to locate. Not only that but I had simply turned up, walked up to the top of the hill and spotted the attention-grabbing poles that had been kindly placed to pinpoint the orchids for visiting admirers. It had been easy. Too easy. Once again, I found myself conflicted; had these poles not been there I may never have found the Small Whites, but on the other hand a large part of the enjoyment I gain from botanising is in the search, and eventual discovery, of what I am hunting for. Often well-meaning botanical tourism can reduce the satisfaction; in this case my encounter with *Pseudorchis* had been something of an anti-climax.

After a quick lunch, I jumped back in the car and drove down to the coast to visit Kenfig National Nature Reserve. The serenity of the meadow in Powys quickly diminished as I journeyed south, the wooded hillsides giving way to bland concrete and the smoke-churning towers of Swansea. Located on the south-eastern edge of Swansea Bay, Kenfig is one of the last remnants of a vast dune system that once extended from the Gower Peninsula along the coast to Bridgend. The dunes are of national significance, spreading inland for nearly two miles and hosting many rare species of plants and animals. Large tidal flats and steady winds have contributed to the growth of the sand dunes and their continued turnover. Kenfig is one of two areas in the country where the Fen Orchid is known to grow– it used to have a population of thousands, but there has been a severe crash in numbers in recent years and now they are lucky to have twenty.

As it turned out, I didn't even need to leave the car park to find my first orchid – not a Fen but a Pyramidal, a glorious pink beacon in the long grass and a contender for the friendliest of our native species. It was Ray again who first recorded what is now known as *Anacamptis pyramidalis* in Britain, as late as 1660, despite the plant's abundance. This widespread and relatively common orchid is the county flower of the Isle of Wight and has always been one of my favourites. More conical than pyramidal when it first appears, the flower spike quickly becomes cylindrical as more flowers open, providing its localities with a wonderful splash of colour. This single flower was my thirtieth species of the year.

Energised by reaching this next milestone, I set out to look for number thirty-one, the elusive Fen Orchid. I was armed with extremely detailed instructions from Suzie Lane, a contact who had been to see the plants two days previously. She had found two individuals, only one of which had been in flower. At first, I had laughed at her seemingly over-the-top directions, but now I realised that there was a reason they had to be so long. I rounded the visitor's centre and was met by a great expanse of sand dunes. Kenfig was a maze, riddled with discreet pathways so inconspicuous that you had to think twice about whether or not they were just animal trails. And somewhere in there were two Fen Orchids. I set off into the dunes, performing a strange trotting walk as I followed Suzie's directions meticulously, knowing that without them I wouldn't stand a chance.

Four hours later, I was standing in the middle of a dune slack with wet, sandy mud sucking at my feet. I was very lost. All around me creeping willow carpeted the floor, its fluffy seeds floating past on the light breeze. The dunes rose up on all sides so that I had absolutely no idea where I was; only the faint sound of the sea in the distance gave me an indication of which way south was. This was not easy.

Having followed numerous dead-end trails, each time convincing myself that I was following the correct one, I had successfully managed to lose all sense of direction. Some trails looped back on themselves, bringing me to a path I had already walked; others started promisingly, only to peter out altogether, leaving me in the middle of nowhere. I was getting more and more confused, overwhelmed by the immensely intricate network of paths and wondering whether I would ever find the right dune slack. It was now 5:30pm; I was running out of time.

I began picking my way across the slack, stopping from time to time to admire the gloriously rich colour of the Early Marsh Orchids that grew scattered among the cottony willow. They were the subspecies *coccinea*, deep red in colour and a dune slack specialist, some squatting close to the ground while others rose phoenix-like from the dark peat. On the far side, I was momentarily distracted by ten Southern Marsh Orchids that were standing in an auditorium of electric-blue viper's-bugloss. A single Common Twayblade sprang from the marram, twisted at a bizarre angle as it attempted to escape the foliage. Honeysuckle and burnet-roses lined the path, and a meadow brown danced up in front of me – the first I had seen this year – and fluttered along, seemingly oblivious to the human presence following in its wake.

According to Jocelyn Brooke, 'the Fen Orchid is one of those species which owes its rarity to the farmer rather than the botanist'. First recorded along with the Pyramidal Orchid in 1660 by John Ray in Cambridgeshire, it is now an Endangered Red Data Book species and included in Schedule 8 of the Wildlife and Countryside Act (1981). Restricted to a mere handful of sites, its stronghold remains in the Norfolk Broads, where subspecies *loeselii* continues to endure despite extensive drainage of the fens and their conversion to agricultural land. More recently, the decline in traditional practices such as peat cutting, grazing and reed harvesting has led to further losses, particularly within protected areas, many of which are now giving way to alder and willow carr, a habitat which offers no place for the under-stated Fen Orchid.

In south Wales, the situation is no better for subspecies *ovata*. While nine sites have been identified over the years, a combination of the over-stabilisation of the dunes and reclamation of land for large-scale industry has resulted in a substantial decline in suitable habitats and, subsequently, population sizes. Kenfig now holds the only viable population but even here it remains feeble. Very few new slacks are formed as the dune system ages, which appears to be the core of the problem. The Fen Orchid is a pioneer species, adapted for colonising dynamic environments; new dune slacks have an abundance of bare ground that can be exploited but new populations are soon crowded out by larger perennials. While turf stripping and grazing can be employed as a short-term remedy, the long-term stability of this orchid depends on the instability of the dune system and the continual generation of young slacks.

The lack of Fen Orchids here was concerning, but a thought had suddenly occurred to me. I turned and ran up the side of a dune, pulling my GPS from my bag. Suzie had sent me a grid reference, not for the plants directly but for the dune slack that they grew in, which I had hurriedly scribbled down the day before at the bottom of my notes. How could I have forgotten that? Waiting irritably for my machine to load, I glanced to my right, where I could see the smoke-filled, industrial skyline of Port Talbot Steelworks in the distance. I wondered what used to grow there before it was built, and whether or not Kenfig and its precious flora and fauna would someday suffer the same fate.

A bleep told me my GPS was ready. I punched in the

numbers and set off in the direction of the arrow that was now guiding me, four hours late. I was moving urgently now, completely ignoring the paths and cutting straight across the dunes. I knew it was a bad idea, but desperation had begun to kick in.

I ran, up and over, up and over, until I finally scaled a particularly large dune with a wide slack in its shadow. The noticeable lack of mature vegetation stood this depression apart from the previous slacks I had traipsed through, an obvious indication that it had been cleared a year or two before. This was the right spot, definitely; it perfectly matched Suzie's description. It was Fen Orchid territory. I tumbled down the hill, almost falling as I skidded onto the sparsely vegetated sandy soil, realising too late that I would be unable to stop myself flattening the one remaining orchid I had come to see if it happened to be growing at this end of the slack. Fortunately, that didn't happen, although the lack of orchids wasn't necessarily a good thing.

Countless meadow thistles and brookweed plants grew in the sea of muddy sand that had been scraped over. Small vegetated islands punctuated the dull brown expanse, providing havens for young willows, lesser spearwort and common restharrow. Around the perimeter, lush vegetation supported thousands of Southern and Early Marsh Orchids, and there were Marsh Helleborines in tight bud. Twenty minutes passed, but still no Fen Orchids.

Once far more common, they were collected at will as publicised by this account of a club outing to Burwell Fen, Cambridgeshire, in 1835: 'We had very good sport in both plants and insects. *Ophrys loeselii* [Fen Orchid] was found in great plenty. Between four hundred and five hundred specimens were brought home. It was growing in the grass and moss among the pits where they cut turf. There were two bulbs to each plant, and the bulbs were scarcely in the ground at all, so that we picked them out easily with our fingers.'

I could have done with 500 plants then. Another twenty minutes went by. Exasperated, I took out Suzie's directions again and tried to work out where she had entered the slack. Perhaps it made more sense from the other end? I hurried over, stepping carefully between thistles and bedstraws. I was distracted momentarily by the sound of vibrating wings, and turned to see a dragonfly buzz past, but then my eyes were drawn to two small sticks, stuck haphazardly in the ground. The tell-tale rare-plant marker.

Relieved, I ran over and bent down, searching, willing a Fen Orchid to appear. It didn't. Disappointment washed over me, followed by more anxiety. But the sticks! What were they marking then? Cruel. And then a bare bit of ground amongst the small grasses and a pair of oval leaves: absolutely minis-cule. *Liparis*, unmistakeably, but no flowers. I took a breath; the plant Suzie saw must be here. She had even said that there were two sticks in the ground. Everything fitted and yet I couldn't find a flower, only more waxy leaves, tiny lime ovals cupped together.

I was panicking now. This was my one and only oppor-tunity to see the Fen Orchid. At such a crucial stage in the season I really didn't have time to make another trip to Wales, and the East Anglian sites are widely considered too fragile for visitors. I sent a worried text to my father before resuming my search, eyes combing over the same area again and again, making sure I hadn't missed anything. My earlier misgivings about the ease with which I had been able to find the Small

Whites suddenly seemed rather ironic. I would do anything for a garish signpost now.

Fifteen minutes later, after I had found perhaps eight little pairs of leaves, I sat up and rearranged my aching limbs, the success of my project hanging by a thread. This wasn't how it was meant to be. I cast a cursory glance around and froze: more sticks. Another two, ten metres away; I scrambled to my feet and charged over to this second patch, heart racing. Here there were more sterile plants and then, a second later, I found one with a flower spike. It rose up from the leaves but then stopped abruptly: the top had been nibbled off.

I sank to my knees, unable to believe what I saw, unwilling to accept that I had failed. The Fen Orchid, flowering only two days ago, had been eaten. Gone; a tasty snack for a passing rabbit. Despair swept through me. This wasn't fair! I had come so close, spent so much time and energy searching and was now faced with the long, empty-handed drive home. The memories of my grand tour would be forever tarnished with the knowledge that I had failed to find a flowering Fen Orchid. I wasn't exactly surprised that it had been the Fen that had stumped me; this orchid is so small, so rare and so camouflaged that the majority of people searching for it end up going home disappointed. A species present in fleeting yet precious moments in the ever-shifting, transient dune system.

I took out my phone and found the photo that Suzie had sent me of the tiny green plant she had found, as if seeing it would magically bring it back. If only I had visited a few days earlier. I paused mid-thought then, realising that the orchid in the photo was surrounded by the yellow buttercup-flowers of lesser spearwort. The nibbled plant in front of me was on its own in the sand with only a few tufts of sedge to keep it company; there was no lesser spearwort. Once again, I felt hope rush back. Vegetated islands with lesser spearwort; the thought ran through my head on repeat as I looked around me. There were lots of suitable contenders. The first was empty, as was the second, save for more brookweed. The third, however, elicited a reaction one would normally associate with a last-minute title-winning goal: 'Yes, yes, yes, get in!' I whooped. Two Fen Orchids, one in bud and the other in perfect flower, stood proudly in the middle of some creeping willow, lesser spearwort sprouting yellow all around them.

Relief, pure relief, was all I felt. I began laughing; this was ridiculous. My mad shouts were whipped away by the wind, leaving me dancing with excitement. To have missed one after thirty species would have been devastating, but here it was, a very satisfying Fen Orchid, complete with a short pale-green spike rising from its pair of buttery oval leaves. The somewhat spidery flowers were inverted, facing the sky, the thread-like petals forming a cross. I couldn't believe I had found it.

Fen Orchids employ several reproductive mechanisms, one of which is self-pollination by rain. A falling raindrop hits the upturned lip, which deflects it onto the anther cap. This in turn is knocked sideward so the pollinia come into contact with the stigma, resulting in pollination. Vegetative reproduction may also occur, whereby 'buds' are formed on the pseudobulb (a thickening of the stem used to store carbohydrates). These detach in the autumn, complete with their very own fungal symbiont. This form of reproduction will often be used in the absence of rain and insect pollinators.

The species' Latin name, *Liparis loeselii*, is derived from the Greek *liparos* meaning 'greasy', referring to the appearance of the leaves; they look waterproof, they're so shiny. The specific name *loeselii* is a tribute to the Prussian botanist Johann Loesel (1607–1657).

I knelt in the mud, still reeling; marvelling at the sight of my Fen Orchid. Taking out my phone, I sent a grateful message to Suzie. Her detailed directions had saved my quest. I knew I wouldn't have found these tiny orchids without her help. I didn't know it then, but she would have a much bigger impact on my summer later down the line.

The satisfaction was immense, all the better for having been tormented by the elusive flower for the majority of the afternoon. I had found the flowering Fens at 7:15pm and had now been in the dunes for more than six adrenaline-pumped hours. Exhaustion was creeping up on me. I was scratched by wayward brambles and covered in a congealing concoction of mud and sand. It was time to call it a day.

By the time I had finished photographing the plants, packed up my things and started heading back to the car, the sun had begun to set. A sea breeze had picked up, making a scratchy rustle in the marram. I squelched back around the slack, boggy trails, which then became sandy paths and finally pebbled walkways. The meadow in Powys seemed an age ago. It had been a wonderfully rewarding day.

My path took me to the edge of the reserve, providing a view out to sea, which was as shiny as the leaves of *Liparis*. The evening light glanced off the wet sand, soon to be part of the dunes as they are formed and reformed. Maybe, just maybe, there might be new slacks to come yet, and with them the instability that is so important for this humble little orchid.

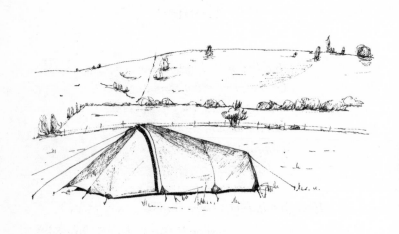

13

Mimic

'And few of that most curious race,
Or those that rival them in grace,
Perhaps exceed; the Ophrys kind
In the advancing season join'd,
Stamp'd with their insect imagery,
Gnat, fly, butterfly and bee,
To lure us in pursuit to rove
Through winding coombe, through shady grove.'

Bishop Richard Mant (1776–1848)

Wiltshire and Kent
June 2013

Orchids are specialists in deception. There are orchids that imitate nectar-producing flowers and entice pollinators by pretending to offer food. Others attract flies by mimicking the appearance and smell of rotting flesh. Orchids of the genus *Serapias* produce flowers that look like burrows for hiding in, while some tropical species pretend their flowers are male bees in flight, provoking territorial fights that result in pollination. Approximately one-third of the 25,000 orchid species have figured out that they can con animals, be it through visual, textural or aromatic deceits, or, in the case of the insect-mimicking *Ophrys* orchids, all three at the same time.

The genus *Ophrys* is famous for its sexually deceptive pollination mechanisms. Of the four species regularly found in Britain, I had already seen two: Early Spider and Fly Orchids. Their lifelike flowers had been deceiving insects throughout the spring. This system is so elaborate, so ingenious and so unthinkably clever that it has led advocates of intelligent design to use orchids as proof of the hand of a creator. It certainly highlights challenging questions about evolution. How did this mechanism become so intricate? Given natural selection's tendency to side with simplicity, why have orchids developed such complex reproductive strategies rather than sticking with more straightforward mechanisms involving a nectar reward like most flowers? And what do the insects gain from this act of deception? What possible advantage could there be for them, in order for this whole act to have been retained by natural selection? Such a specific strategy is vulnerable to stochasticity, as many orchids are so specific they depend solely on an individual insect species: if this pollinator dies out, what does the orchid do then?

It has taken years, but evolutionary biologists have provided answers to all these questions. Pseudocopulation, to give it its scientific term, wasn't observed until 1916, more than fifty years after Darwin published *The Various Contrivances by Which Orchids Are Fertilised by Insects*. Darwin was frustrated by the seemingly purposeless extravagance of insect-mimicking orchids observed on a grassy bank near Cudham in Kent – popularly known as Darwin's Orchis Bank. It puzzled him for years. It didn't fit with his theories about evolution. In 1829, the Reverend Gerard E. Smith wrote: 'The Bee-Ophrys has, indeed, the appearance of that insect, engaged in pilfering a flower; Mr Price has frequently witnessed attacks made upon the plants by a Bee, similar to those of the troublesome *Apis muscorum*; and I have myself seen a young entomologist, approaching stealthily, with outstretched hand, the successful deceiver, whose mimic beauty became, alas! its own ruin.'

Darwin had never seen any insect visit the Bee Orchid, which has abandoned the traditional mechanism in favour of self-pollination, but was puzzled as to why Mr Price's bee was attacking the flower. Where did this little plant fit in his theory of natural selection? How had natural selection shaped a flower that resembled an insect so successfully that it could fool an entomologist?

Fast forward 150 years, and we know that orchids are more devious and cunning than Darwin ever suspected. The bee described by Smith was not attacking the flower, but trying to mate with it. Research carried out on *Ophrys speculum* by Maurice-Alexandre Pouyanne, a French colonial judge in Algiers, helped shed light on Darwin's conundrum. Once the insect lands on the orchid, Pouyanne noticed that 'the abdomen tip becomes agitated against these hairs with messy, almost convulsive movements, and the insect wiggles around'. While investigating this behaviour, he realised that all the visiting insects were male and that those messy, convulsive movements only occurred in the few weeks before the females hatched. After conducting numerous experiments in order to establish the source of the attraction, Pouyanne concluded that the insects' movements were analogous to those of insects attempting copulation.

A few years later, an Australian naturalist by the name of Edith Coleman independently came up with a similar idea. Upon discovering the works of Pouyanne and a British naturalist, Colonel Masters John Godfery, who had confirmed the Frenchman's results, she synthesised a more detailed account of this phenomenon. She was the first to conclusively prove these ideas by finding wasp semen on her flowers. These three botanists had independently arrived at the same conclusion: that insects mistook the orchid flowers for females of their species and were trying to have sex with them. The orchids had evolved to exploit the short period of time between the emergence of the males and the emergence of the females.

So why have these strategies developed? Why move away from offering a nectar reward? There have been studies showing that if the pollinator doesn't immediately find what it's looking for it thrashes about in search of the food or mate it's sure is there, in doing so increasing the odds that it'll unwittingly pick up the pollen. Other studies suggest that it's to do with dispersal. When nectar was added to orchids that normally didn't provide it, the pollinators began hanging around. They would visit other flowers on the same plant, and other plants in the vicinity. This results in inbreeding, which is ideally avoided as it decreases the health of its offspring. The frustration of the pollinator is key in this system, encouraging the bee to leave quickly and fly a fair distance before trying another plant. It favours mixing genes with far-flung individuals which are less likely to be closely related. Having such a specific sexually motivated strategy, targeting one or two pollinator species might seem unstable, but actually there may be benefits. Nectar attracts everyone and anyone, meaning a lot of your pollen will probably be delivered to the wrong species. By attracting a very specific species, you improve the odds that your pollen will be delivered to your own kind.

Interestingly, the Bee Orchid, *Ophrys apifera*, has long since given up trying to lure insects to its flowers. Instead, it is routinely self-pollinated, despite clearly having evolved to attract male bees as pollinators. Botanists have puzzled over why this might be. The most likely scenario is that sometime in the fairly recent past, the particular bee that pollinated the Bee Orchid became extinct. In order to survive, the Bee had to make a dramatic lifestyle change. Another possibility is that the pollinator switched from feeding on the Bee Orchid to another plant, perhaps a different *Ophrys* species. In any case, this switch doesn't seem to have affected its welfare, as it remains widespread in England and across Europe. Inbreeding leaves it vulnerable, though, and the effects of climate change may well have a disastrous impact on Bee Orchid populations in the future.

Despite its aura of rarity, the Bee Orchid is actually quite easy to find and is often discovered in new locations. Most regularly occurring in short-turf pasture and chalk downland, it is associated with earthworks and old quarries. Man-made habitats are particularly popular, such as railway embankments, roadsides and industrial estates. These locations are perhaps the last places you would expect to find orchids.

I had chosen to search for the Bee Orchid in the grounds of Salisbury District Hospital. In my hand, I was clutching the hastily scribbled directions received from the woman I had met while looking for Burnt Orchids three weeks previously. People take great pride and feel very protective of their local Bee Orchids. For those with a passing interest in wildlife, the discovery of a small cluster of these plants is an exciting occasion. She claimed to have seen more than sixty plants there last year, so I figured it was worth a shot. Established Bee Orchid populations are notorious for the dramatic fluctuations in the number of flowering plants from year to year, so there was a chance I might not find any. Unlike the Lady's Slipper, the Bee Orchid is very short-lived.

It felt odd to be walking into the grounds of a hospital on a warm Friday afternoon in search of orchids. There were long queues of cars arriving for visiting hours, interspersed at intervals by ambulances. Although they are relatively common, I had only seen eight Bee Orchids in my life, and all of them had been growing in old, species-rich grassland.

Reaching the entrance, I glanced down at my crumpled sheet of paper. *Walk into the reception area and head straight down the corridor in front of you, taking a left after the double doors.* This was the strangest orchid hunt I had ever been on. Doing as instructed, I turned the corner, dodging doctors and nurses as they hurried past. *At the end of the corridor turn left again, then right, before passing through two doors.* I complied, suddenly realising that I was following these instructions blindly, not stopping to check whether I was actually allowed there. No sooner had the thought entered my mind, I felt a hand on my shoulder.

'Excuse me, are you lost? You're not allowed through there,' said the doctor. I had just been about to pass through a third door. Apologising profusely, I backed up and felt her watching me suspiciously as I walked away. Paying closer attention to the directions, I followed them all the way through to a second reception area on the other side of the hospital. I walked back out into the glaring sunshine. A large sign by the door read 'Welcome to Salisbury District Hospital'. This was clearly the main entrance.

Still wondering why I'd been sent on a journey through the labyrinth of hospital corridors rather than directly to the front door, I spotted a steep grassy bank in between two of the car parks: probably a good place to start. Within ten seconds, I had found my first Bee Orchid. One minute later, I had clocked at least twenty.

Consumed by childlike curiosity, I sat down next to a group of seven plants clustered around the bottom of a lamppost. At the centre of three bubblegum-pink sepals, the lip sits velvety and bee-like. Its furry surface is a rich chocolate brown, overlain with a mix of irregular yellow lines and splodges. An inner purplish-brown loop outlines a base of cinnamon. The other two petals are rolled into long lime antennae either side of the green column which rises from the flower, culminating in what can only be described as a duck's head. This is where the yellow pollinia are produced and I could see some of them hanging down and swinging gently in the wind. Some had already stuck to the stigma: self-pollination in action.

The plant's English and Latin names clearly both refer to

its resemblance to a bee. The specific name *apifera* is derived from the Latin *apis*, meaning 'bee', and *fero*, meaning 'to bear' or 'to carry'; together they mean 'bee-bearing'. European names also follow this trend: in France it is '*Ophrys abeille*', in Germany *Bienen Ragwurz*, or bee ragweed, and in Italy *Vesparia*, meaning 'wasp'. In Britain, local names include Bee-flower in Wiltshire, Honey-flower in Kent and Bumble-bee in Devon and Somerset. In Dorset, it is known as the Humble-bee Flower and in Surrey it is a Dumble Dor. John Gerard, who was the first to record this species in 1597, calls it the Humble Bee Orchis. Humble, I thought, was a fitting adjective.

I moved from Bee to Bee, taking in all the intricate differences between each flower that made them unique. I found one which had unrolled a little golden stinger at the tip, but instead of looking threatening it reminded me of our cat Tabitha when she forgets some of her tongue is sticking out. Some were much paler than others, with pastel-pink sepals. I was struck by how happy each flower looked. This was definitely Britain's friendliest orchid.

As it was visiting time at the hospital, there was a steady stream of people walking past me. By now I was immune to the stares and mutterings of passers-by, but I was surprised no one asked what I was doing. I wondered whether any patients would be hearing about the boy on the bank.

I felt my phone buzz in my pocket. It was a text from Dom Price with exciting news. He had been walking around Figsbury Ring that morning with his family when his wife, Susie, had found a Bee Orchid on the ramparts. I almost dropped my phone in excitement. I had looked for them at Figsbury every summer since I had first found one there in 2001, but had never had any luck. How appropriate, I thought, that it had chosen this summer to reappear.

I dropped in at Figsbury on my way home. Once considered an Iron Age hillfort, Figsbury Ring is now thought to have been a Neolithic henge. Instead of serving as a defensive structure, it would have been used for ritualistic purposes, like Stonehenge, which isn't far away. People still process around the circular earthwork rings, children and dogs running along the chalky path, though unobservant of any sort of ritual.

There was just a single Bee growing on the raised earthwork amid hundreds of Chalk Fragrant Orchids and a cloud of gentian-blue milkwort. It was spectacular, somehow far brighter and more handsome than the plants up at the hospital. This one had two bees, resting peacefully on candyfloss-pink flowers. Their patterned flowers were more yellow than brown, and I wondered whether they might be a named variety.

There are six named varieties of Bee Orchid: individuals with distinct patterns that occur frequently enough to warrant nomenclatural recognition. One, called *chlorantha*, is ghostly pale with white sepals and a greenish-yellow lip. *Bicolor* has a plain yellow petal that looks like it's been dipped in melted chocolate, and *trollii*, also known as the Wasp Orchid, has a narrow lip that tapers to a sharp point. The Bee in front of me looked like it could be a *belgarum*, with a horizontal band of yellow across the middle.

The Bee Orchid is one of the surprisingly few British species that has inspired its own poem. Two stanzas by Langhorne describe these little bees settled in front of me:

See, on that flow'ret's velvet breast

How close the busy vagrant lies!
His thin-wrought plume, his downy breast,
The ambrosial gold that swells his thighs.

Perhaps his fragrant load may bind
His limbs;– we'll set the captive free –
I sought the living bee to find,
And found the picture of a bee.

Two days later, I woke up in a very hot tent. Outside, I could hear the voice of a boy talking urgently into a walky-talky: "We need to move *now*. I'll meet you by the washing area. The girls are in the woods, but we don't have much time. Over." I rolled over. It sounded too risky to go outside just yet.

I was back in Kent, at the very south-eastern tip of Britain, and had come to find two of Britain's coolest flowers: Lizard and Late Spider Orchids. I had never seen either of these species and was unimaginably excited.

Crawling out of the tent five minutes later, I took in the bright-blue, cloudless sky and the morning sun flooding the campsite. Down below, the English Channel glittered invitingly. It was going to be another beautiful day and, according to the weather forecast, the hottest of the year so far.

Piling my tent into the back of the car, I trundled out of the campsite and drove the eight miles to Sandwich Bay. It was 10:30am and I was full of anticipation as I set off down a narrow dusty path to the sea. Sandwich Bay is a largely inactive dune system and is home to several rare species. It is perhaps most famous for holding the largest Lizard Orchid population in the country. Curiously, they grow all over the Royal St George golf course.

The footpath led me through the dunes and onto the golf course. A sign reading 'Please be aware of golfers approaching from the right' seemed to have missed the point as a small white ball went whizzing past.

Pausing to let a couple of mid-morning golfers tee off, I spotted a small tortoiseshell on the path in front of me. Its wings were a palette of orange and yellow with a border of blue sapphires. I approached carefully, but disturbed it. It flicked up and swooped twice around me before disappearing over the marram.

I passed through a corridor of viper's-bugloss, an electric-blue guard of honour. By the path grew several broomrapes of varying sizes. Broomrapes are odd-looking plants that are parasitic on the roots of other species. These were bedstraw broomrape, parasitic on bright-yellow lady's bedstraw, and smelled wonderfully sweet, of clover. This little plant is incredibly rare and, it turned out, can only be found here on the east coast of Kent. Not a bad rarity to stumble across. Further on there were more good finds: pin-striped funnels of sea bindweed, juicy-leaved biting stonecrop and some Pyramidal Orchids standing out like beacons in the grass.

The Lizard Orchid was first recorded in 1641 in *Mercurius botanicus* by Thomas Johnson. He found it 'nigh the highway betweene Crayford and Dartford in Kent'. From the 1600s right through until about 1850, it was restricted to this area and, due to habitat destruction and over-collecting, by the turn of the nineteenth century the Lizard had been reduced to just four sites. But then things took an unprecedented turn. In the subsequent thirty years, it dramatically expanded its range, establishing thirty-six populations as far west as Devon and north to Yorkshire. Being on the edge of its range in England, it appears that small changes in climate have a significant impact on its distribution. Fortunately, the Sandwich Bay population seems to have thrived under protection. Because of the threat posed by collectors, this colony was under twenty-four-hour guard during the 1970s and 1980s.

A couple of yachts, triangular smudges on the water, had ventured around the headland from Margate and were making their way along the horizon. The warm air was being bandied about by a light breeze and had picked up the clove-like scent of the bedstraw broomrape. Suddenly though, as the wind changed direction, I was hit by a revolting, pungent farmyard animal smell that made me wrinkle my nose in disgust. What on earth was that? Glancing to my left, I got my answer: Lizard Orchids.

Holding my breath, I darted over and knelt down to examine these bizarre flowers. Lizard Orchids are unmistakeable. Their robust spikes of greyish-green flowers are commonly about thirty centimetres tall and at first sight appear crowded, ragged and untidy. Hooded by pale-green sepals, the lip is long and quickly fades from white to maroon-brown as it corkscrews through two or three turns, ending with a notch at the tip some seven or eight centimetres away from the plant. It supposedly resembles the tail and hindquarters of a lizard,: however the front half appears to have been swallowed by the flower. When inside the bud, this 'tail' is coiled up like a spring. At the base of the lip, where it is still white, there are small pink dots which resemble little faces.

In many ways far from the beauty and fragility one associates with the word orchid, the Lizard is a heavyweight. Exploring the dunes, I found many that exceeded sixty centimetres. A handful of enormous gangly spikes down by the road were all a metre tall. These plants were hipsters, with their scraggly beards of brown, coiled lizard tails. In *The Wild Orchids of Britain*, Jocelyn Brooke describes the whole plant as having 'a distinctly uncanny and sinister air, and were it more familiar, one would not be surprised to see it forming part of the Freudian dream-patterns of the surrealist painter'.

The smell was making me feel sick. Anne Pratt was quick to label it 'perfectly disgusting', while a friend of Jocelyn Brooke's who served in the First World War claimed that the plant smelled like the battlefield. To most people they smell of billy goat. The species Latin name, *Himantoglossum hircinum*, makes reference to the strap-like lip (*himas* is 'strap-like' and *glossa* means 'tongue', giving us *Himantoglossum*) and to the goat-like smell (*hircinus* means 'of a goat'). Interestingly, the French name for this species, *Orchis bouc*, is a further reference to goats, while in Kent it is locally known as the Great Goat-stones. The Goat Orchid, though, doesn't have the same ring to it, nor does it conjure up quite as exotic an image.

In *Wildflowers of Chalk and Limestone*, Ted Lousley gives an account of his friend, John Jacob, who became completely obsessed with Lizard Orchids in the late nineteenth century. 'It is not surprising that such an erratic and bizarre plant has held an irresistible fascination for at least one Man of Kent. My friend John Jacob of Dover made the quest for "Lizzies" his life work from the time he saw his first in 1885. For twenty-five consecutive years of those that followed he never missed a season without seeing the plant somewhere or other. The list of localities he wrote out for me before his death is a long one. There can be little doubt that he held the record for having seen the greatest number of "Lizzies" in England.'

John Jacob's talent for tracking down Lizard Orchids was unrivalled. One outing with Lousley in 1925 resulted in them discovering just how bad its scent was. 'Earlier that year,' Lousley writes, 'two elderly ladies had found what they took to be a wild Aspidistra; they dug it up and kept it in a pot at their home. Towards the end of June it came into flower, and they then realised that they had found something very much out of the ordinary and tried to get the plant named. Their inquiries came to Jacob's notice and he called on the ladies and collected the plant with the object of placing it on show in Dover Museum. We took the plant on the train together and during the journey gained practical experience of a feature of the Lizard Orchid which was new even to Jacob. In the ordinary way in the open air in daytime the most unpleasant smell given off by the flowers is not noticeable [I beg to differ]. On this occasion it proved to be exceedingly objectionable, and the reminiscence of the he-goat implied by the scientific name *hircinum* was shown to be well founded. Our journey took us through lengthy tunnels in which the carriage windows were closed in accordance with custom, and on this warm summer day the smell was overpowering. If the ladies who dug up this rare orchid in ignorance had kept it in their house they would have met with a just reward!'

Thankful not to be in a confined space with these plants, though not convinced Lousley had ever experienced them outdoors, I reached the other side of the golf course. I was amazed by the sheer volume of Lizard Orchids. Despite its rarity, the Lizard is an adaptable species and has often been recorded in strange places: there are regular records of it popping up in people's front lawns and there is a well-known colony spread across the racecourse at Newmarket. It's almost as if it got fed up of waiting to be found in remote locations.

I began following the road along the beach. Sea-kale and common mallow grew in tidy, organised islands in the shingle. I spotted the blue-green plastic leaves of sea-holly among yellow rattle and seeding sea sandwort and lanky wild asparagus growing like spindly trees.

The road was lined with Lizard Orchids. Some were short and stunted hidden in the long grass, while others were enormous towers of grey, green and various hues of brown; my favourites were a dark chocolate burgundy. I found drooping lizards with tails twisting like fusilli, plants that wagged their tails in the wind like dogs, and some whose tails looked more like the unrolled proboscis of a moth or butterfly. One of the highlights of the day was coming across a spike that had its 'ribbon-like streamers', as C. B. Tahourdin calls them, stuck out on end like hair filled with static electricity. A Mr G. Chichester Oxenden described the Lizard as the 'Monarch of Orchids' and provided the specimens used by Darwin in his description of the species in his book on the fertilisation of orchids. They were extremely special plants and I was utterly enchanted.

Several hours later, having dragged myself away from the Lizards, I found myself high up on the chalk hills of Kent.

The time had come to look for Late Spider Orchids, the fourth and final insect-imitating *Ophrys* species. Its specific name *fuciflora* literally means 'deceitful flower'.

Late Spiders are unthinkably rare. Restricted to Kent, their small populations are limited to fewer than ten sites, only five of which will reliably produce flowering spikes every year. Around 50 per cent of the British population grow near Wye, east of Ashford, and droves of orchid tourists flock there every June to admire them through their wire cage prisons. Although the enclosures are constructed to protect against rabbits rather than humans, I didn't consider a hill-side covered in mesh cages a fitting introduction to one of the coolest orchids in the country, so I decided to track it down elsewhere.

I had driven through Dover and parked up on the downs with a picture-postcard view out over a sparkling sea. The hillside below was a long stretch of steep, rolling fields. Small hollows had been carved out from the slopes where landslips had occurred, leaving snowy drifts of chalk. The result was a downland full of secret pockets, all calling out to be explored.

Remarkably, the first British record of a Late Spider Orchid dates from 1828, when it was noted 'on the southern declivities of chalky downs near Folkestone' by the same Reverend Gerard E. Smith whose observations on the interaction between bee and Bee Orchid had so puzzled Darwin. Given its rarity, it is hardly surprising this species wasn't recorded until the nineteenth century. In fact, it is possible that an ameliorating climate helped a seedling establish itself after being blown over from the continent, and it has been here ever since. This might explain why it is restricted to this tiny corner of Kent.

Five minutes along the path and I was already deep into one of the clefts incised into the hillside. It was full of common wildflowers: fairy flax, horseshoe vetch, yellow-wort and mouse-ear hawkweed and dotted with Common Spotted Orchids. In the remains of an archaeological settlement, I found Pyramidal Orchids and a few Bee Orchids, their friendly faces smiling up at me.

Thirty minutes later and I was still searching. Conscious of the long journey home, I decided to give up and try elsewhere. As I trudged back to the car, exhaustion creeping up on me, I was surprised to stumble across a cluster of Man Orchids. They were well past their best, but nevertheless provided a happy flashback to my evening on Darland Banks. I hadn't expected to see them again so soon.

And then I spotted the Late Spider Orchids: crisp and intensely coloured. Late Spiders are more closely related to Bee Orchids than Early Spiders and this is evident in the flowers. The oval sepals were the same candyfloss-pink, with a prominent green line running down the middle. However, the Spider's two non-specialised petals were different from the Bee's: small, dark pink and furry. They looked like horns, some the deep red of strontium flames. The spider's lip was squarer than the Bee's, and a rich chestnut brown with a golden tip. Each one had a different Aztec-like symbol emblazoned across its back. It was inherently obvious why their velvety surfaces were so attractive to insects.

Unlike the Bee's, the Late Spiders were feisty, almost aggressive in their showmanship. They're rare and they knew it, as if they would immediately challenge anyone or anything in the vicinity. Unfortunately, they're becoming even rarer. The Late Spider is pollinated by insects, but their pollinators

– certain solitary bees in the genus *Eucera* – no longer exist in the UK, meaning very few flowers are pollinated. This begs the question: how do they survive in Britain? Self-pollination may occur occasionally, or pollination by other insects like pollen-beetles, although this has never been observed. It seems that individual plants are long-lived.

One of my former biology teachers, Paul Collins, noted that the long-horned bee, *Eucera longicornis*, has been recorded occasionally along the south coast of England, presumably mostly a result of strong south winds carrying it over the English Channel. He found this highly amusing, commenting: 'I bet the males that do make it are frustrated as hell by the lack of females to mate with and are very grateful to find some extremely rare bee porn.' I think he was probably right.

Who knows what the future holds for the Late Spider Orchid? Will climate change devastate their tiny populations in this corner of Kent? Or will they adapt and start attracting some other helpless insect species? Even now, they could be following in the Bee Orchid's footsteps and slowly moving towards self-pollination. Only time will tell, but I hope they stick around; I have seldom come across such an inventive plant.

To appreciate the intricate, deceptive charm of the *Ophrys* orchids is to admire one of evolution's finest products. Natural selection, through random chance, has come up with a system that allows a plant to exploit one of an animal's most urgent needs: sex. But it makes you wonder who else orchids have duped. Have we, who assume superior intelligence, actually been tricked into becoming arguably the most important orchid pollinators in the world? Propagating them, continuing genetic lines and passing on precious genes to the next generation on their slow but purposeful march towards botanical domination? Our actions have allowed orchids to exploit a range of new habitats: supermarkets, florists, perhaps your living room windowsill. We've spent all this time mocking the intelligence of a handful of hapless insects when it is perhaps us who are the real fools.

14

Queen of the Cotswolds

*'But the queen of this part of the Cotswolds is
the Red Helleborine'*

J. E. Lousley, *Wildflowers of Chalk and Limestone* (1950)

Gloucestershire
July 2013

The Red Helleborine is on the brink of extinction in Britain.
After the success of the Lady's Slipper reintroduction programme, this orchid is the rarest in the country, save perhaps
the Ghost. Yet unlike these two, it is rarely talked of and
seldom searched for.

The first record appeared in *English Botany* in 1797, which
stated that it was 'gathered last June on Hampton Common,
Gloucestershire, by Mrs Smith, of Barnham House in that
neighbourhood'. For a long time, the Cotswolds remained the
species' UK stronghold. In *Wildflowers of Chalk and Limestone*, Ted Lousley claimed 'records for the Red Helleborine,

spread over the last century and a half, extend over a great many of the beechwoods from Nailsworth to Birdlip, but it has seldom been seen in quantity and flowering is extremely erratic'.

Many records and herbarium specimens, including orchids, were assembled by members of the clergy. Vicars who had simple rural parishes, a lot of time on their hands and a love for natural history were perfectly placed to observe and document the local flora. It was an ideal pastime: outdoors, at one with nature, while engaging with the parishioners as they wandered down the lane. Writing in 1948, the Reverend H. J. Riddelsdell documents the diminishing number of sites in the Cotswolds as botanical collectors heard of the Red Helleborine's presence: he 'saw some fifty or sixty plants together, but only about ten bore spikes of flowers, and somebody cut those before the next morning'. In Gloucestershire, it is currently present in only one known site.

In 1955, three flowering spikes of Red Helleborine were discovered in a wood in the Chiltern Hills of Buckinghamshire. They flowered intermittently and then disappeared in the 1970s, but were found again in the 1980s.

Then a third population was discovered in 1986, this time in Hampshire. Other than a dubious record in 1926, this was a completely new location for the Red Helleborine. It was thought that felling of trees in the area a few years before had stimulated this appearance, and so more felling has been carried out in an attempt to encourage the plant further.

As it is a priority species, a Biodiversity Action Plan (BAP) has been drawn up to keep it from extinction. In the UK, it is now classified as Critically Endangered under the IUCN Red List criteria, while in Europe it is Vulnerable.

Local conservation is implemented in the UK at each of the three sites by Natural England, the Berks, Bucks & Oxon Wildlife Trust (BBOWT), and Hampshire County Council.

In 2005, the entire known British population consisted of sixteen plants, less than half of which produced flower spikes. The situation was critical. In order to stop the decline of the plant in the UK, the Red Helleborine Restoration Group was formed, to share knowledge of the plant's requirements. There were representatives from Natural England, the National Trust and the Royal Botanic Gardens, Kew.

Surprisingly, the three populations that remain contain a significant amount of genetic diversity, meaning it might be possible to restore this species to sustainable numbers. For reintroductions to occur, we must be able to grow the Red Helleborine from seed. Efforts in the lab at Kew Gardens have not been fruitful so far though, and with no success there, instigating a reintroduction programme would be challenging. The Restoration Group is currently looking at French sites where this orchid grows in good numbers to try and determine its exact growing requirements.

To make things more difficult, the Red Helleborine is extremely high-maintenance, even for an orchid. It's incredibly sensitive to grazing and light levels so site protection and management are crucial. Given the size of British populations, any management has to be handled very tentatively: one wrong move and the population will disappear.

I had been desperate to see the Red Helleborine for years, agonising over photos on the internet every summer. The site in Hampshire was close to where I lived and so would seem the obvious target. Unfortunately, though, they haven't had a flowering plant there since 2004 and in the last five years they haven't had any vegetative plants either. That one looked like a dead end. The Gloucestershire site is effectively off-limits, as the plants there are enclosed within a high metal fence and it is not open to the public. If I wanted a photo, this wasn't an option. In the Chilterns, however, the plants are more accessible. While this site is also within an enclosure, people can obtain permits from BBOWT and join organised pre-booked visits to see and photograph the plants – I'd signed up for one during the winter in anticipation of the coming July.

On the 21st June, two weeks before the Red Helleborine was likely to flower, I received an email from BBOWT. I opened it excitedly, expecting to read details about the visit. Instead, my heart sank:

'It is with regret that I contact you with the news that the Red Helleborines are not going to flower this year and consequently we will not be issuing any permits for guided visits. The Warden informs me that this is the first time since 1988 that no flowers have been produced. It is very difficult to be certain why this has happened, particularly as many other plants seem to be doing well, if somewhat late. The Warden feels that the probable cause is the lack of sunshine, a factor that does seem to be important for their full development. I would like to offer my apologies for being the bearer of disappointing news.'

I fell onto my bed. For whatever reason, this summer had not been quite good enough for the Red Helleborine and so it had refused to come out to play. First Hampshire had been a dead end, and now so was Buckinghamshire. That left the site in Gloucestershire, an impenetrable compound in the woods. Admittedly, you could see the plants from the public footpath that passes the perimeter fence, but I would need binoculars and wouldn't be getting anywhere near close enough for photos. It was a huge blow: this rare and beautiful orchid was one of the species I had been most looking forward to seeing.

Over the next few days, I wallowed in frustration, trying to work out a way to see the Red Helleborine. I was desperate. But eventually I had an idea. I called Dom Price. In his role as director of the Species Recovery Trust, he might have some sort of link with the conservation of the orchids. This charity focuses on the species closest to extinction in the UK.

With perhaps the greatest turn of fortune so far, it turned out Dom played a small role in the Red Helleborine Restoration Group. I couldn't believe my luck. I told him about the distinct lack of floral activity in Buckinghamshire and then voiced my hopeful plea: was there any way he could get me into the Gloucestershire enclosure?

I waited nervously for three days. Dom made phone calls and sent emails asking whether it would be all right for me to be shown the plants.

I was at home when I received the text from Dom:

Red Helleborines are on!

The Cotswold hills begin near Bath and curve north-east past Gloucester, extending almost as far as Banbury in Oxfordshire. It is an Area of Natural Beauty, or AONB, thirty miles across in places. Most of the best sites for plants, and scenery, lie near Gloucester and Stroud, where steep escarpments provide lots of unique habitats as well as extraordinary views.

I drove up to Gloucestershire to meet Timothy Jenkins, the National Trust ranger who oversees the plants. We were joined by his colleague from Natural England, Kate, who had come along to help Tim survey the plants.

We jumped into the back of a huge National Trust 4x4 that stood outside the farmhouse-style offices on the estate and bumped our way down an old dirt track. Suddenly, we were in the wood. My first thought was how enclosed it felt: it was haunted-house dark. To our left the ground fell away and I could see the rough surface of the tree canopy down below. It was a long way down were anything to go wrong. I gripped my seat a little tighter as the truck rattled and rocked. Tall beech trees covered the slope, and delicately filtered the sunlight so that it seemed to float down from the canopy far above. I spotted three ghostly Greater Butterfly Orchids lurking among the vegetation.

We continued deeper into the wood, always moving down the hill, but twisting and turning until eventually we came to an abrupt halt next to a tall wire fence.

The enclosure was a lot smaller than I had imagined: it was the size of someone's back yard. When told I would need binoculars to see the plants from outside, I had envisioned a much larger compound. The fence was at least two metres tall and topped with barbed wire. It looked ominously impenetrable.

I was desperately excited, craning my neck to see whether I could spot the first helleborines. Tim took out a key and unlocked the padlock, swinging open the large door so that we could all shuffle inside.

Never have I been more terrified of accidentally stepping on a plant. I followed Tim down the hill, taking care to step exactly where he did. What if I stepped on an orchid and made him regret ever listening to Dom?

The enclosure was a jungle of plants. The ground was littered with saplings: ash, beech and sycamore all struggling to thrust themselves towards a gap in the canopy. In among some wild strawberry floated the white jellyfish flowers of common wintergreen. There were yellow hawkweeds, bramble and dog's-mercury everywhere, all leaning towards the light. This didn't seem like the home of a rare orchid.

Yet just below us, poking out of the foliage, was a fragile pink spike. Against all the odds, a Red Helleborine.

It is a humble orchid, silently stunning, possessing a spindly stem topped with an elegant rose-pink flower spike. Each inflorescence seemed too big for the plant to support. It was a miracle they could hold themselves upright. Any small breath of wind instigated a drunken swaying. The sepals were fuzzy with small hairs. It was unimaginably delicate. I got the impression it didn't like the attention and would be quite happy to just be left alone. If this orchid could speak, I'm not convinced it would say much. In *Wildflowers of Chalk and Limestone*, Lousley writes: 'But the queen of this part of the Cotswolds is the Red Helleborine, *Cephalanthera rubra*. It is one of the most elusive of British orchids and many wildflower hunters have spent a surprising amount of time and money in its quest. One of my friends even received a cable from a lady in Monte Carlo who was anxious to fly over to see it.' Strange, since it's far commoner in France.

Tim and Kate took out clipboards and began counting the plants, each marked by a pale kebab skewer. There were a lot of non-flowering spikes which were incredibly difficult to spot in the thick vegetation. I was completely engrossed in my photography, aware that I wouldn't have a second chance at this.

Natural pollination is rare in British Red Helleborines so most of it is done by hand. Interestingly, recent work in Sweden suggests that Red Helleborines and certain species of bellflower share the same pollinators, namely small solitary bees of the genus *Chelostoma*. Red Helleborines don't produce nectar. The bees are attracted to the orchids, because of the colour. As they aren't sensitive to the red end of the spectrum, the flowers appear pale-blueish lilac, like certain bellflower species that do produce nectar. This pollination mechanism might be a clever swindle, except the absence of these *Chelostoma* bees from Britain means that natural pollination by insects is virtually non-existent.

Most of the plants were numbered, but I overheard Tim saying 'Bella's going to flower, four buds'. I asked him why it was called Bella. Apparently, these plants are so difficult to spot that the initial survey fails to pick all of them up. Any plant that is found subsequently gets named after the person who spotted it. Tim's daughter, Bella, had seen this plant from his shoulders. I was impressed.

Upon learning this, I immediately ducked down and began searching the surrounding vegetation for unmarked helleborines. It would be seriously cool to have a Red Helleborine named after me. Twice I thought I had found one, only to spot the kebab skewer seconds later. The leaves were so narrow and the plants so small that I was sure there must be at least one that they had missed, but after twenty minutes of fruitless searching, it was time to go.

We trundled back up the warren of muddy tracks to the top of the hill, leaving the enclosure and its precious contents in the depths of the woods. I began looking through the photos on my camera and, to my dismay, realised that the plants had been in bud. Extraordinarily, blinded by excitement, I had failed to notice. I returned a week later and observed the helleborines, in full flower, through my binoculars from the public footpath that ran past the bottom of the enclosure, my camera hanging uselessly by my side.

The Red Helleborine is hanging on in Britain. For how much longer it is difficult to tell, but the hope remains that this highly sensitive species could suddenly pop up and flower elsewhere in the country. Such is the furtive, capricious, enigmatic world of orchids.

15

Orchids of the Western Isles

'The handsomest and most interesting flowers were the
great purple orchises, rising ever and anon, with their
great purple spikes perfectly erect, amid the shrubs and
grasses of the shore.'

Henry David Thoreau, *The Maine Woods* (1864)

Yorkshire and North Uist
July 2013

The moorland was bathed in the purple of midsummer
heather. Beside me, a liquorice-coloured stream ran down the
hill, its dark peat banks lined with waxy rushes and cushions
of moss. Appearing from nowhere, a short-eared owl ghosted
across the moor, impervious to the buffeting wind, and made
its way towards me, flapping silently.

Turning my head to follow its flight, I picked out the fiery
pink flowers of rosebay willowherb and suddenly I was ten
years old again, playing hide-and-seek among the boulders

that lay scattered on the hilltop. My family and I spent hours here every summer. My sisters and I used to play in the sandstone quarries or help pick bilberries on the steep slopes to take home for pudding. We would return with ice cream tubs full to the brim, our mouths and fingers stained with tell-tale purple.

I was back in Haworth, taking a short break before continuing my journey north to Scotland. After a morning inside reading, I was making the most of the afternoon sunshine. I began wandering down the track towards the road. Fine dust billowed up in clouds as I walked, the consequence of a string of hot, dry days. To my left the stream flowed slowly: brown water seeped through boggy flushes and trickled over miniature waterfalls formed by stones lodged in the peat. The grass beside it was pebbled with sheep's droppings. The fluffy heads of cotton-grass, bobbing like rabbit's tails in the breeze, mingled with sneezewort, meadowsweet and tiny yellow tormentil.

As I walked, meadow pipits flitted along the dry-stone wall, always one step ahead of me. A battered wooden signpost directed me onwards: Brontë Falls 2 ½ miles. The sandy track extended, and across the valley, stone walls created a higgledy-piggledy patchwork of green fields, with small houses dotted here and there.

Reaching the end of the track, I turned onto the road and started walking back down the hill towards Haworth. As I descended, I began noticing orchids growing in the damp grass. They were Northern Marsh Orchids, my thirty-sixth species of the year.

Of all the marsh orchids, this species is fairly easy to identify: the lip of the flower forms a deep-magenta diamond and is scrawled with red loops and swirls. Most plants have a tightly packed inflorescence and are richly coloured. No other marsh orchid has the dark crimson overtones of the Northern Marsh. The leaves on these plants were unmarked and a bright, polished green.

As the name suggests, this species has a northerly distribution. It is commonly found growing in marshy grassland across the north of England, Wales and Ireland, and is a regular feature of Scottish meadows. Interestingly, it has very little overlap with the Southern Marsh Orchid, which is the dominant marsh orchid in the south. There is a small band running diagonally upwards from Wales to south Yorkshire where the two species meet and occasionally hybridise. Curiously, it was one of the later additions to the British orchid flora: it wasn't recognised as an individual species until it was described from a sighting of plants near Aberystwyth in 1920.

The roadside was cluttered with rich spikes of Northern Marsh Orchid. I stepped gingerly through the colony. I couldn't linger: I had to continue north, up into Scotland and west to the Outer Hebrides. I was on the hunt for the Hebridean Marsh Orchid, one of the most localised species in the British Isles.

The next two days were spent driving through Scotland. I camped on the shores of Loch Lomond, passed through Crianlarich and Fort William, and stopped on the roadside far more than I had planned in order to admire Scotland's wonderful scenery. As I proceeded, the mountains rose up around me, the no-man's-land between them washed lilac with heather. Lakes and tarns were pooled across the landscape. Ben Nevis was still peaked with snow.

As I turned west towards the Hebrides, the sun seemed to fade, shrouded by wisps of cloud, before disappearing altogether. It began to rain as I reached Skye. I flicked my lights and windscreen wipers on. Forty minutes later, the rain had set in and was coming down in drizzly sheets. I had pulled into a small layby up on the moors and was shivering as the air temperature plummeted. The prospect of camping in this weather was depressing.

When I arrived at the small village of Uig, the cloud was so low that I could only see ten metres ahead. I bumped up the pot-holed track, dispersing puddles of cloudy water. I sat in the car, willing the weather to turn, but it was miserable and so was I.

As I got out of the car, the mud squelched thick and brown under my feet. There weren't many campers about, but there were several caravans lined up against the fence. I gazed enviously at the warm glow coming from the one nearest me. I unpacked my tent, still damp from the previous night, and began pegging it out, using the car as a windshield. Midges swarmed around my head. After wolfing down some pasta, I retired to the car. I was cold and lonely and sat there longing for company.

I called home and whimpered to my parents, who tried their best to console me. It was twenty degrees warmer at home. England was experiencing a heatwave and my family had spent the afternoon sending photos of ice cream, sunbathing and G&Ts. At that moment, I couldn't have cared less about orchids: I wanted to be at home, warm and dry.

Over the next hour or so, I watched as the rain died away and the cloud began to shift. The wall of greyish white that had surrounded the campsite was fading and I could just about make out silhouettes beyond: dark, shapeless masses and moving lights. Above me, a small Scottish flag rippled aggressively in the wind as if struggling to free itself from its rusty post.

Eventually, as the mist cleared, I realised I was looking out over the sea. The lights I had seen were coming from a bullish ferry that was now sitting in the harbour, humming loudly as it prepared for its final journey of the day.

As I stood there, the last layer of cloud was sucked away and sunlight, pure and bright, streamed down over the large hills that backed onto the campsite. Boats moored to the jetty were now brilliantly coloured: red, yellow, orange and blue. Down on the shore, the sun's rays penetrated deep into the water and transformed a brown mass of seaweed into glinting shapes of green, ochre and auburn that swirled with the slow current.

I was freezing and couldn't feel my toes. Across the harbour, I could hear a gentle clinking as rigging met mast. I watched as the ferry chugged past, dragging itself out into the open water and setting its course for the Outer Hebrides. First thing in the morning, it would return and I would board for the last leg of my journey to North Uist.

Waves rolled in lazily over the pale-blue sea, crashing onto a stark-white beach. The bleached sand seemed to stretch for miles and curved out and round as it followed the arc of the bay. In the distance, the islands of Lewis and Harris were dark smudges on the horizon. The hill, awash with clover and buttercups, sloped gently down to the sand dunes which rose and fell as far as the eye could see. Grey-green marram grass covered the slopes. Behind the dunes was an extensive area of low-lying grassland. Atlantic storms blow sand and fragments of shells over the dunes and onto the coastal grassland beyond. The result is short, species-rich turf. This habitat is known as machair and is unique to western Scotland and parts of Ireland.

It had gone midday by the time I arrived at Clachan Shannda, a tiny village in the north of the island. I had driven down to a small cemetery and parked on the grass, then followed a stony track that disappeared enticingly over the hill. I was now standing at the top of that hill, thrilled by the beauty of the Hebrides. The landscape before me felt raw and untouched.

I skipped down the slope in gleeful anticipation of the botanising ahead and took a narrow sandy path behind the dunes in the direction of the Machair Walk. Progress was slow as there were so many plants to look at: common stork's-bill, sand sedge, wild pansies and tiny blue bugloss. The small white flowers of thyme-leaved sandwort twinkled at me from the darker sand and Northern Marsh Orchids spilled across the path. I checked each one, hoping to find the Hebridean speciality I had come for. I was completely in my element and lost all track of time.

The Hebridean Marsh Orchid is an enigma. It was first discovered in 1936 by M. S. Campbell, and was originally thought to belong to the amalgamation called Western Marsh Orchid (this included what is now the Irish Marsh Orchid).

During the latter half of the twentieth century, it went through names like wildfire, until it was eventually granted full species status, as *Dactylorhiza ebudensis*. After I had drawn up the species list for my trip, it underwent another change: a demotion from species to variety. It is currently known as *Dactylorhiza traunsteinerioides* ssp. *francis-drucei* var. *ebudensis*, or in English: a kind of Pugsley's Marsh Orchid. Dom Price pointed out there should be a rule against Latin names this long and he hoped I found it faster than it took me to say it. As a result of this name change, I had been in a dilemma over whether or not I would come to the Outer Hebrides to see this orchid at all. But after deliberating, I decided that as it had been a full species when I was planning my trip, it would remain on my list of species to track down.

Back on the machair, I was startled by the loud squeak of a lapwing overhead. It swooped down low before shooting skywards again, mobbing me. The further I walked, the more frequent these warning calls became until I was subject to constant attention from three or four birds. I was obviously getting perilously close to their nests. The correct thing to have done would have been to turn back and walk away, but that would have meant leaving the machair and I hadn't travelled all this way to be nice to a few lapwings.

Keeping an eye out for nests, I tried my best to ignore the high-pitched squeals from the sky. I looked around me and grinned: the machair was dotted with hundreds of marsh orchids, and not just purple ones, but pink, red, white and every shade in between. There were Early Marsh Orchids everywhere: both the brick-red dune slack specialist *coccinea*, and the flesh-coloured subspecies *incarnata*. In some cases, it became difficult to tell which was which. One spike looked like an ice lolly, pale pink at the base and slowly getting darker at the top. The identity of that one was anyone's guess.

The genus *Dactylorhiza*, to which all the marsh orchids belong, hybridises for fun. I had seen a few hybrids on my travels, but most of them had been rare and easy to identify: Lady x Monkey Orchids and Sword-leaved x White Helleborines. The parent species of those hybrids had been present and the plants were clearly intermediate between the two. The marsh orchid hybrids, on the other hand, don't follow either of these rules. In most cases, only one parent species is present, and in many cases neither parent can be found. The hybrid itself often looks nothing like either parent. In such a large colony of marsh orchids, like the one here, hybridisation was inevitable.

That such extensive interspecific breeding is possible is down to genetics. The marsh orchids are all very closely related as, on a geological timescale, they only split into separate species very recently. While they remain genetically similar, they can hybridise. Over thousands of years, however, small changes to their DNA will accumulate, and the species will become increasingly different. Eventually, long after you and I are gone, the species will be so different that they will no longer be sexually compatible – and therefore won't produce hybrids. It gets more complicated, though, as fertile hybrids can themselves become new species. The Hebridean Marsh Orchid is thought to be the result of a long-ago hybridisation event between Early Marsh and Common Spotted Orchids. Whether a hybrid becomes a new species or not is dependent on a long list of environmental and genetic factors, too tedious to rehearse.

I thought the plants in front of me were probably a cross

between Northern Marsh and Heath Spotted Orchids. While the former was present in numbers, the latter was nowhere to be seen. The grassland was rich in calcium: a Heath Spotted Orchid's worst nightmare. Its absence didn't rule it out, though. Perhaps, I wondered, it could be the hybrid between Northern Marsh and Common Spotted? There were certainly Common Spotted Orchids around, although only a few. They were a special Hebridean form with more heavily pigmented petals and accentuated concentric looping patterns.

It was while trying to come to a conclusion that I laid eyes upon my first few Hebridean Marsh Orchids. I had driven more than 700 miles to see this dumpy purple plant. It looked remarkably similar to the Northern Marsh Orchid, but was more stunted and its lip was three-lobed rather than fashioned into a diamond shape. It was also a deep, royal purple, unlike the reddish purple of Northern Marsh, and its leaves were invariably splotched with large uneven patches of dark brown. The most distinguishing feature was that all the flowers on the inflorescence faced the same way. From one angle, the plant looked like it was in full flower; from the opposite side it appeared to lack flowers completely. This was a bizarre phenomenon that I had also observed in the Dense-flowered Orchid and Pugsley's Marsh Orchid in Ireland.

Delighted by my find and relieved to have found it relatively quickly, I continued to explore the machair. A trio of Early Marsh Orchids stopped me in my tracks and I bent down clumsily to photograph them. I have a knack of getting into awkward and painful positions while taking photos, too absorbed in the moment to care.

Taking stock, I stood to relieve my aching limbs. The fields beyond were misty with clover. I had been watching a herd of highland cattle drift around their pasture throughout the afternoon. They were close enough to the fence for me to hear the quiet snuffling as they tugged at the turf. In the distance, there were three white cottages with slanting brown roofs and old red doors. The hills rose up behind them, their surfaces pocked with slivers of exposed granite.

I knelt down to look at some marsh arrowgrass near a small stream and suddenly spotted a single Frog Orchid growing in the sandy soil. It was so small and green that even my trained eye had almost missed it. It came as a surprise: I hadn't considered the possibility I might find one up here.

Varying in colour from mostly green to very red, Frog Orchids tuck their frog flowers out of sight, camouflaging themselves among the surrounding vegetation. Their flowers are bent backwards so far that they are frequently more or less horizontal. A certain amount of imagination is required in order to see the animal in the flower, but the lip is said to resemble a frog in full hop. Even if this is beyond the observer, certain parallels can be drawn with their colour and introverted nature. The sepals were reddish and hooded the frog, trying desperately to keep it concealed. Another, more mundane name for this plant is the Bracted Green Orchid: efficient and to the point, but rather unimaginative.

Despite looking nothing like marsh orchids, genetic analysis has shown that Frog Orchids are also part of the genus *Dactylorhiza*. This makes sense when you consider the frequency with which the Frog Orchid hybridises with other species of *Dactylorhiza*. The hybrid with Common Spotted Orchid has been widely recorded, while it has also been known to cross with Heath Spotted and, very rarely, Northern Marsh Orchids.

The Frog was first recorded in 1650 by William How in his *Phytologia Britannica*. John Ray also included it in his flora of Cambridgeshire in 1660, and six years later, Christopher Merrett wrote of it growing near Oxford and Lewes. It has a widespread, if patchy, distribution across the UK and is generally commoner in the north, particularly Scotland. Strangely, it is completely absent from Kent and only has a few records from pre-1970.

My phone was buzzing away in my pocket: Andy Murray was attempting to win his first Wimbledon title and I was missing it. My father was texting me regular updates. Unfortunately, the phone signal was patchy at best, so every twenty minutes I received a barrage of messages:

30-15
40-15
game to Murray!!

I was desperate to watch the match.

After a picnic lunch, I crossed over the stream, stopping only to have a look at the sea-milkwort that grew in some dried-out ruts. There were more Frog Orchids here and an impressive collection of Early Marsh Orchid subspecies. A flock of waders whizzed past. My father would have been able to identify them, I was sure, but I was too out of touch with this group of birds. Plovers, perhaps? The squealing lapwings were giving me a headache.

North Uist was wild. Truly wild. Not the 'wild' we are promised by nature reserves, but the original, untouched 'wild' associated with the extreme Scottish islands. The only signs of civilisation were the tiny ramshackle huts. The nearest city was a very long way away. Instead of silence, the island was filled with the sound of nature: insects chirping, wind rustling the marram, and even those pesky lapwings. Finally, orchid hunting at its best.

The serenity of the machair and surrounding landscape was deeply calming. I found a warm, sandy hollow halfway up the dunes and sat gazing out to sea. The small, uninhabited island of Lingeigh guarded the bay, its rugged shores beckoning invitingly. The sea was a pale turquoise, lightened by the white sand underneath. The sun, which had been battling the clouds all day, had just begun to seep through the haze and was quickly warming the air. All around me insects were buzzing and hopping. One lilac tufted vetch flower next to me had five black-and-red burnet moths clustered on it, all enjoying the plentiful nectar and completely oblivious to the satisfying click of my camera. More texts suddenly burst through.

30-30
30-40
Murray has broken Djokovic!

I walked back through the dunes, discovering more Frog Orchids of amazing diversity in the scrubbier areas between the sandy bunkers. There were dark-red ones that stood twenty centimetres from the ground, alongside a tiny five-centimetre-tall plant that was incredibly green, save for the lip of each flower, which was chocolate brown.

Slipping through a bank of marram and sea rocket, I jumped down to the beach and walked along the tideline, awestruck at the expanse of brilliant white sand. At the end of the beach, where sand met shingle, I pottered among the stones and boulders. North Uist is primarily made up of banded Lewisian gneiss and pinkish granite. Some of the rocks here are three billion years old. On a whim, I picked up a fragment and tucked it into my pocket: a small memento of my afternoon on the machair.

Murray's two sets up.
Nail biter.

Turning to leave, I caught a glimpse of an oystercatcher floating over the sea. It passed behind a rocky crag and emerged again, the sunlight playing across its dark back. I followed it as it sailed across the horizon before it disappeared behind the hill, its cries suddenly muffled and then gone altogether.

I had been distracted by something as I followed the wader's flight. I glanced around, trying to work out what I'd seen. There in the grass: a purple orchid. Its flower spike was longer than that of the marsh orchids but the colour not nearly so royal. Each lip was extended, drawn out. I couldn't quite believe it, but I had found the Frog x Northern Marsh Orchid hybrid. I jumped around, surprising myself with my excitement. This was a ridiculously rare plant.

My only sadness was that there was no one around with whom to share my find. I was leaping around, but there wasn't even a passing stranger to whom I could show it. Once again, I felt the longing for an equally enthusiastic friend to accompany me on trips to see plants, but I knew it was extremely unlikely to ever happen. The number of young people interested in plants is dwindling rapidly. Interest in botany in this country is at an all-time low. In 2013, Bristol University waved goodbye to the final handful of students on its undergraduate botany course, the last of its kind in Britain.

The disappearance of botany from UK universities – a remarkable landmark, surely, which seems to have passed unnoticed – is symbolic of a shift in focus from taxonomy and classification to genetics and molecular biology. It is now possible to complete a plant science course in Britain without once identifying a British wildflower. The withdrawal of our final botany degree epitomises the fact that the need to identify British plants is no longer deemed important or relevant in a society rapidly losing interest in the plant kingdom.

Casting this dispiriting thought aside, I fetched my stove and some food from the car, passing the oystercatcher now stood on a gravestone in the old cemetery. I climbed one of the dunes and gasped at the view: this second beach, called Tràigh Hòrnais, swung round like a boomerang, nestled between the dunes and the sea. The high-tide line was marked by a dark band of seaweed: kelps sprawled helplessly, fronds like ganglia; a loose strand of thongweed had draped itself over a hunk of driftwood. It was a perfect spot.

I found a sheltered hollow in the marram, set my stove up and began cooking my dinner. Closing my eyes, I listened to the rhythmic, metronomic swash and backwash of the sea and the cry of the gulls. It was by far the wildest, most tranquil place I had been so far and I loved it. Checking my phone, I had one final text:

Murray's won Wimbledon!

16

Midsummer Musk

*'The orchids have been called the Royal Family among
flowering plants: a happy comparison, though hardly
flattering to human royalty, if one considers some of the
tropical orchids, with their debauched, pendulous lips and
unhealthy mottled complexions. Nor, for that matter, on
strictly physiognomic grounds, is it particularly flattering
to the orchids themselves.'*

Jocelyn Brooke, *The Wild Orchids of Britain* (1950)

Hampshire and Conwy
July 2013

That week in Hampshire I saw my first Musk Orchids of the
summer. With the success of my trip to Scotland edging me
closer to forty species for the year, I had returned, once again,
to the chalk. My drive south had brought me into the thick
of the heat wave: England was sweltering.

I had received emails from a couple of locals in Hampshire telling me of Musk Orchids in flower at Noar Hill. I had seen them there several years previously: my mother and I had visited one sunny July evening at the start of the summer holiday and discovered swarms of them.

I had been orchid-hunting alone for far too long, so I rang up my friend Sam and invited him along. Sam is a keen naturalist and wildlife photographer and probably the closest I've come to finding a botanical friend. He had just returned from his first year studying in Abu Dhabi and was full of stories as we crossed the border into Hampshire.

Noar Hill is a Hampshire Wildlife Trust nature reserve near Selborne. It is the site of ancient chalk workings, abandoned in Medieval times and left to wander into the welcoming arms of nature. The chalk pits form sheltered bowls on the hilltop that are filled with plants and insects. Ancient deciduous woodland borders the summit, encouraging a mosaic of habitats. The reserve boasts thirty-five butterfly species, including the Duke of Burgundy and brown hairstreak, and eleven different orchids.

We arrived at Noar Hill at five o'clock. It was still extremely hot and we were both sweating by the time we reached the top of the hill. Against the clear blue sky, the shimmering haze blurred the chalky earthworks into an optical illusion. The grassland was awash with colour: yellow lady's bedstraw, the purples of tufted vetch and knapweed, and white ox-eye daisies standing tall in the grass.

It took a couple of minutes to reach the edge of the Musk Orchid colony. Estimated to be 10,000 strong, this is one of the largest orchid populations in the country. So I didn't find the first one: I found the first twenty. Musk Orchids are small lollypops with delicate yellow-green bell-like flowers, each ending in three little prongs. There's something fairy-like about them. Instead of musk, they actually emit a sweet scent that has been likened to honey, although it was very faint on this hot afternoon.

Musk Orchid is exclusive to well-drained, short grassland on limestone soils and has a preference for chalk. It is confined to the south of the British Isles and absent from lots of seemingly suitable habitat. Unlikely to be stumbled upon, this very local species requires pre-planning.

The colony at Noar Hill has built up quickly, far faster than most orchids. They are pollinated by small flies and beetles, but they also use vegetative propagation as a way of reproducing. This means that the root system produces extra tubers up to twenty centimetres away from the parent plant which then grow their own flowering spikes. Over time, the direct connection between plants is severed, forming individual genetic replicas. The result: a vast clone army of Musk Orchids.

Once again, it was John Ray who came up with the first published record of Musk Orchids in Britain in 1663. He found '*Orchis pusilla odorata...* in the chalk pit close at Cherry Hinton'. Its Latin name, *Herminium monorchis*, is thought to derive from *Hermes*, the messenger of the gods. While *Hermes* is known as a trickster in some myths, the Musk is an honest, by-the-books orchid that diligently provides its pollinators with the nectar it promises. Alternatively, its name could come from the Greek for 'buttress' which would supposedly describe the single pillar-like tuber found in this genus. However, Musk Orchid has a spherical tuber – *monorchis* means 'one testicle'.

We wandered between battalions of the clone army on Noar Hill, stopping for twenty minutes at a time to take photos in companionable silence. There were more Pyramidal Orchids here than I had ever seen before, standing out like bright pink lychees among the dry grass. And with a lot of variation: some were so pale they were almost white and had a wavy, nearly solid lip. Others were vivid cerise and deeply lobed, and had pronounced ridges at the opening of the spur. These are thought to guide insects into the flower so that they are in the correct position for pollination, and can access the sugary sap inside the spur. Once the mechanism has been triggered, a small spring-like structure called the viscidium attaches itself to the proboscis of the visiting moth or butterfly, so that it can be transported to another flower.

It was a shame Sam hadn't been around this summer as it was far more fun being with someone else. He was interested in everything: orchids, butterflies, birds. He was the one friend who never thought to make fun of my interests, perhaps because he shared so many of them.

While inspecting some of the weirder Pyramidal Orchids, we were interrupted by a small, bumbling man who introduced himself as James. He said he'd been watching us and wondering what we were doing. After we'd talked for a few minutes, his eyes suddenly lit up and he asked us whether we also liked butterflies. The answer, of course, was yes from both of us.

James tapped his nose, winked and asked whether we would like to see a Duke of Burgundy. Sam and I looked hesitantly at each other. I'm often approached by naturalists in the field, characters of all sorts, from quiet and cautious through to bouncy and exuberant. Whichever category they fall into, they are invariably delighted to meet a fellow enthusiast and keen to share what they have found. The Duke of

Burgundy is found on Noar Hill, but it was far too late in the year for this little orange-speckled butterfly. I didn't have the heart to tell James this, though, as he was clearly extremely excited by it. Putting any concerns aside, we followed him over to the top of a bank about ten metres away where he stopped, put his fingers to his lips and stood stock still. We waited, not entirely sure what to expect.

'There's one!' he declared, pointing to a large meadow brown struggling to escape from a prison of long grass. Sam threw me a questioning look, eyes narrowed. 'Look, there's another,' cried James. This time it was a ringlet. Neither Sam nor I could bring ourselves to tell James that these butterflies weren't the rare Duke, merely common everyday species that can be found across the country.

After a further five minutes, during which James politely enquired as to why 'two lads like us weren't down the pub', he bid us farewell. On our way back to the car, we passed an information board highlighting species to look out for and I saw at once why James had been confused. The images were beyond identifiable: there was a solid pink triangle on a green stick labelled 'Pyramidal Orchid' and a brown splodge with a few orange dots labelled 'Duke of Burgundy'. No wonder James had been misled.

As I've mentioned, for a long time, I had wished for a friend who shared my interest and enthusiasm for botany. It had always seemed like a long shot, particularly in the twenty-first century. Botany is not considered cool and to admit you like it is to surrender to much mocking.

Earlier in the year, I had been given instructions via email to find the Fen Orchids at Kenfig by Suzie Lane. We had struck up a friendship, messaging back and forth ever since, discussing orchid sites and which species we wanted to see next. She was twenty-two and I quickly learned that her interest in wildlife was broad. She was an ornithologist at heart, but had recently discovered orchids. We talked for several weeks. Towards the end of July, I needed her help again, so invited her to accompany me in search of Dark Red Helleborines in north Wales.

I pulled into the car park at Chester station half an hour early and sat there feeling unexpectedly nervous. I tried reading but found that I couldn't concentrate. What if we didn't get on in real life? I'd tried not to build this up in my head as a big deal but I inevitably had; and I didn't want to mess this up. I checked the time: still twenty-five minutes to wait. What if she arrived early? I scanned the crowd waiting outside the station entrance.

Twenty agonising minutes later, I got out of the car and walked over to the entrance to wait. I checked my phone: no messages. Then I felt a tap on my shoulder and there she was. She was tall with wavy auburn hair and a large camera bag slung over her shoulder. We exchanged awkward greetings and began walking back to my car. She had a Mancunian accent and I was immediately very aware of how posh I must sound to her.

Five minutes into our journey, any initial tension had passed. It turned out we had been to many of the same places and before long we were sharing our strangest orchid-hunting experiences. It was interesting to hear about her search for Creeping Lady's-tresses in Cumbria and Fen Orchids in

Wales. She was annoyed that she had arrived too early to see the Fens in full flower. We talked non-stop and were soon driving through Llandudno, with the Great Orme towering above us.

The Great Orme is a large limestone headland protruding into the Irish Sea. Its name derives from the Old Norse word for sea serpent. The steep cliffs and abandoned Bronze Age copper mines provide suitable habitat for a rich flora that includes numerous rare plants. The surface of the peninsula is well known for its limestone pavements, a favourite habitat for the Dark Red Helleborine.

We drove up the one-way road that undulated around the headland. Cable cars slid silently overhead. Suzie had been in charge of obtaining instructions for the location of the Dark Red Helleborines and I laughed as she read them out: 'Pass the limestone pavement, park in a layby one hundred metres on and begin looking for a lone tree by the side of the road.' It was all so familiar: the directions of a botanist.

The Dark Red Helleborine is a member of the genus *Epipactis* and is the first of what orchid hunters like to call 'true helleborines'. *Epipactis* was first used as a plant name by Theophrastus (*c.* 370–285 BC) and derives from the Greek word *epipaktoun*, meaning 'close together', supposedly referring to the placement of the sepals. The number of species in Europe is debated, with estimates ranging from a conservative fifteen to a seemingly excessive fifty-four. Here in Britain, there are eight species, all of which flower in the latter half of the summer. The *Cephalanthera* helleborines, which flower earlier in the year, are closely related but are thought to have diverged from the 'true helleborines' earlier in evolutionary time.

Epipactis flowers all have a similar structure. The three sepals and two petals are very similar but the lip is divided into two sections. The inner section forms a small bowl-like structure called the hypochile, which is where the nectar is secreted. The outer section, or epichile, is a triangular landing pad with its tip bent underneath the flower. They are delicately constructed flowers, and in the Dark Red Helleborine they are stained blood red.

Having found the layby in the instructions, we parked and walked down the narrow road. A wall of pale stone rose cliff-like on our left; to our right, there was a steep drop down to the sea. In the distance we could see the Welsh coast as it curved back round to the west and over to Anglesey, which was lit up in the sun.

We caught sight of a silver-studded blue bouncing around in the grass in a sheltered spot by the road. We followed it, willing it to land. It eventually did, gripping a blade of grass tightly as it shifted in the slight breeze. Casting our eyes around, we realised this was a small colony and counted at least forty individuals, mostly females, but a few fresh, shiny males. They were the subspecies *caernensis*, Suzie told me, which is restricted to the Orme. I had seen the other, commoner type of this small butterfly in the New Forest.

Suzie suddenly yelped: she had found the Dark Red Helleborines growing by the side of the road at the base of the cliff. There weren't many, ten perhaps, and I was alarmed to find that most of them had almost finished flowering. Maybe their proximity to the coast had accelerated their flowering season? We were also towards the southern extent of their range in the UK, so they were likely to flower earlier.

Dark Red Helleborines are tall and thin. Two ranks of

dusky-green oval leaves give rise to a long stem adorned with striking wine-red flowers. The pollinia and anthers are a bright, contrasting yellow. We took it in turns to photograph the one good plant, already competing with one another for the best photos, while the other acted as traffic warden.

Records for Dark Red Helleborines date back to 1650 when William How noted it in his *Phytologia Britannica* on behalf of a Mr Heaton. John Ray documented it in 1670, growing in Yorkshire 'near Malham, four miles from Settle, in great plenty'. The grykes in the exposed limestone pavement above Malham Cove are a perfect helleborine hideout.

Walking back, Suzie noticed another helleborine growing on a small ledge about three metres above the ground. Frustratingly, it was in perfect flower, far better than the plants by the roadside. I was impressed with this royal-red orchid. It gave off the air of a supercar: shiny, polished and desirable. It taunted us from above.

Back in the car, Suzie was telling me something about a trip to Kent, but I wasn't listening. My brain was working quickly, trying to figure out a way to get at the Dark Red Helleborine on the ledge. I'd had an idea, but wasn't sure if it would work. Dropping the handbrake, I let the car roll slowly down the hill. As we drew level with the orchid, I pulled the car to the edge of the road so that it was as close as possible to the cliff-face.

After assuring Suzie she wouldn't go through the roof, I instructed her to climb onto the top of the car. I stood on the other side of the road, keeping an eye out for traffic. She placed one foot on the roof and the metal sank menacingly. Maybe this wasn't the best idea. But when she lifted her foot, the roof sprang back into shape. She cast a nervous glance in my direction and I encouraged her with a nod. She inched forward. Once in the right position, she straightened up until the orchid was at eye level. It was, we agreed, the nicest Dark Red Helleborine either of us had ever seen.

We got some interesting looks from passers-by: some were amused, others amazed, and all of them were probably wondering what on earth we were doing. Sometimes it takes a little improvisation to get the perfect view of a Dark Red Helleborine.

This had brought me to forty species for the summer so we celebrated with ice creams and a round of mini-golf on top of the Orme. It had been a great day. As we moved round the pitch-and-putt, we began discussing future orchid trips we could do together. After all, I still had twelve species to find. Just to be talking plans made me happy. For me, this new friendship represented the coming together of the two parts of my life: the strange orchid-obsessed me and the me who actually had some normal friends. Could it really be that I was no longer on my own?

Small, drab and unassuming aren't necessarily adjectives you would expect to see applied to members of the orchid family. Many British orchids are considered beautiful, and as a consequence have suffered at the hands of collectors. The tiny green Bog Orchid, though, while rare, has generally evaded the attention of the plant lover. Being the smallest British orchid, rarely growing taller than seven or eight centimetres, it is extremely difficult to find. Add to this the fact that it grows well out of reach in swamps and bogs, and you can begin to understand why. Jocelyn Brooke, writing in *The Wild Orchids of Britain*, remarks that its habitat 'is a further discouragement to any but the enthusiast; and even the enthusiast may hesitate to wade through acres of peat-bog unless he has good reason to suppose that *Malaxis* is to be found in the locality – which, too often, is not the case'.

Malaxis is the former Latin name for the Bog Orchid, which is now known as *Hammarbya paludosa*. The genus *Hammarbya* pays tribute to the small summer residence owned by the Swedish botanist Carl Linnaeus in the eighteenth century. He would visit the Hammarby estate when he wanted to escape from life in Uppsala. The specific epithet *paludosa* simply means 'of marshes'.

A few days after my trip to north Wales, I drove down to the New Forest to try and achieve what I had failed to accomplish with my father all those years ago: find a Bog Orchid. It was mid-afternoon and the warm summer sun had lit up the plains and heathland. New Forest ponies crowded the roadside, slowing the traffic down to a crawl. There were lots of cyclists on the roads, enjoying the weather, and plenty of ice cream vans plying their trade.

I parked the car in a sandy car park just west of Brockenhurst and slipped into my wellies. The bog here was enormous but certainly looked promising: large blankets of *Sphagnum* moss provided the perfect habitat for the Bog Orchid. In 1898, A. D. Webster wrote of the Bog Orchid 'growing on trembling sods of bog-moss, or sphagnum, [where] it almost dares one to venture with impunity'.

I started at the edge of the bog where the golden starlike flowers of bog asphodel sprung from tufts of bog-myrtle. I found white-beak sedge and dangly bog sedge and cross-leaved heath with pink inflorescences like bunches of grapes. By this point, I was seriously regretting wearing my wellies. The heat was rapidly becoming unbearable so I began to edge my way further into the bog, carefully stepping on the grassy islands to avoid causing too much damage to this fragile ecosystem. Pastel-blue keeled skimmer dragonflies buzzed in and out of view, occasionally embracing in vicious aerial battles above me while the daintier large red damselflies quietly went about their business at ground level.

Everywhere I looked, there were red-green sundews: their leaves are hairy ping-pong paddles covered in glistening globules of a sweet, sticky liquid. Looking closer, I could see insects and small spiders trapped in the gluey dew, some struggling against it but only succeeding in aggravating their predicament.

Peat bogs are highly acidic environments. This doesn't mean they are dangerous to us, but to a plant they present some significant challenges. The water contains very little oxygen, and few nutrients. To survive in these tough conditions, sundews have devised a rather gruesome yet crafty coping mechanism: they eat flies. Or rather, they suffocate and consume any insect they can get their leaves on. These hapless critters in front of me were about to be rolled up, asphyxiated and digested by a substance secreted by the plant. What's left would be absorbed through the sundew's leaves.

Across the bog, the birch trees shimmered as a light breeze passed through. Below them, grassy hummocks and bilberry bushes crowded around the perimeter of the marshy areas. I picked a few berries for sustenance.

The smell of a peat bog is unmistakeable: rich, earthy and damp. Close up, it becomes a whole other world. Bending down, I saw sedges, spike-rushes and grey-green, flowerless heathers. *Sphagnum* moss formed rafts of auburn, lime and russet that bubbled and gurgled. The stagnant surface of the brown water was covered in an oily iridescence, caused by iron salts in the peat.

An hour passed and I was still searching. Other than flea sedge, whose mature fruits jump when touched, I hadn't found a thing. Certainly no Bog Orchids. I decided to give up and try somewhere else, so packed up my camera and drove north to Lyndhurst and the bog I had visited with my father.

I was beginning to have doubts about how long it might take me to find this diminutive species. It seemed like an impossible task: find a tiny green plant in a haystack of a bog.

I was greeted by more keeled skimmers, and ducked as one flew straight at my face. It looped round and settled on a sedge leaf, and sat there cleaning its eyes and face, its head swivelling mechanically. Each wing was a perfect, colourless stained-glass window: a net of black veins forming triangles, hexagons and ellipses. It held them horizontally like a biplane; panes of membrane scattering the light. As I approached, it leapt up with an audible flick and was gone. Dragonflies are so difficult to admire up close.

Out on the water, the flowers of insectivorous lesser bladderwort stuck up like periscopes. Instead of trapping prey with sticky glands like the sundew, the bladderwort uses small vacuum-creating pouches that float on the water. When an insect passes the entrance, it triggers a mechanism that breaks the vacuum, causing it to be sucked into the plant, where it is dissolved and consumed.

After fifteen minutes of searching, I found my first Bog Orchid perched on a cushion of *Sphagnum*. It's a strange plant: tiny and green with a single pair of cupped leaves not unlike those of the Fen Orchid. The inflorescence is a narrow cylinder of flowers that resemble miniature cartoon space rockets. Unusually, they look like they're upside down: the lip is at the top of the flower rather than at the bottom as in all the other species. This appears odd at first, but actually these flowers are the right way up, and all the other species' flowers are, in fact, upside down. As the orchid flower bud develops, the lip is at the top, as in the Bog Orchid. In most cases, though, the flower stem twists through 180 degrees so that the lip ends up at the bottom when it opens. For a while it was thought that the Bog Orchid didn't undergo this 'floral resupination' but it does, only it doesn't stop where it is meant to and continues full circle, twisting through 360 degrees.

Before getting any closer to *Hammarbya*, it is wise to carefully inspect the surrounding area to avoid treading on others. In doing so, I quickly spotted two more close by. They were ridiculously small: one was only about a finger nail's length and bore a single flower.

I carefully got down on my knees and smiled as a Bog Orchid came into focus. There was a constant sucking, bubbling and gurgling from the bog as the smell of wet peat filled my nostrils. I was slowly sinking into the mud. Every now and then a raft spider would scuttle across my calf, making me jump.

Despite its diminutive nature and drab colour, I quite liked the Bog Orchid. Perhaps it was the satisfaction of finding this unpretentious plant after searching for so long. In England, there are a few scattered records, mainly centred in the New Forest and Cumbria. There are some outposts in Wales, but other than that you have to go to Scotland to have any chance of finding them. Given its size, habitat and camouflaged colouration, it could easily be under-recorded.

To walkers along the road, I must have looked somewhat eccentric, creeping about in the bog, but they ignored my Gerald Durrell-like figure, in a very English way. Better to just accept and move on than to ask questions. It occurred to me that I was becoming progressively weirder and wilder as the summer went on.

17

Holy Helleborines

*'It is a piece of weakness and folly merely to value things
because of their distance from the place
where we are born, thus men have travelled far enough in
the search of foreign plants and animals, and yet continue
strangers to those produced in
their own natural climate.'*

Martin Martin, *A Late Voyage to St Kilda* (1698)

Wrexham, Northumberland and Cumbria
July 2013

I Love My Job: Leif Bersweden, 19, Orchid Hunter. That
was the title of my interview with the Royal Horticultural
Society. They had contacted me regarding their new initia-
tive to dispel myths about young people and plants, and had
asked if I would be willing to be interviewed as part of a
series involving people under the age of thirty interested in
botany and horticulture.

The interview was being conducted at the RHS Flower Show at Tatton Park. I was nervous as I drove up the M6 towards Manchester, different scenarios looping through my mind. I tried to compose answers to the questions I would most likely be asked and talked out loud to an invisible interviewer.

I pulled into the car park moments before I had agreed to meet the organisers. Unfortunately, in my panic at being late, I had parked at the opposite end to the show and had to run across the site, not entirely sure where I was going.

I arrived half an hour late, out of breath and sweating, spluttering my apologies to anyone and everyone. Before I knew what was happening, a microphone was being attached to my collar and someone was reeling off a list of instructions about what was going to happen. Then, suddenly, I was all on my own. As the camera panned towards me, I felt a bead of sweat trickle down my back. My forehead was shining. Stephanie asked me a question and my mind went blank, sucked into the depths of the lens.

Eventually I started to speak, and became aware of a crowd forming around me as people started listening in. Maybe they thought I was famous. I stumbled my way through questions, gulping for air: I hadn't realised how breathless I was. I could hear my monotonous voice droning on. My throat felt dry and scratchy. Didn't they have some water around here? They filmed me looking at the flowers neatly lined up for the show, which was ironic, given none of them were wild or native.

After an hour of takes, during which the crowd listened intently to my babbling, they were finally satisfied with the footage they had managed to compile and I was free to go. Sitting in the hot car as I queued to leave, I replayed everything I'd said over and over, cringing at how pathetic I'd sounded. I was desperately embarrassed and keen to escape.

I had decided to make use of the trip north to visit some local orchid sites and had arranged to meet Suzie for the afternoon. I collected her from the station and drove across the border into Wales. We picked up where we had left off and I was immediately comfortable again as we laughed about some of my ridiculous interview answers.

We were hoping to find a large population of Dune Helleborine, a species neither of us had seen before. This orchid is a special one, because Britain is the only place in the world where it is found. It is an endemic. Being a small island far from the equator and the tropics, Britain has very few endemic species, so the Dune Helleborine is a big deal. It has had a long and complicated taxonomic history. First granted full species status in 1926, it was then relegated to being a subspecies of Narrow-lipped Helleborine, an unusual orchid of beech woods in the south. After much toing-and-froing, DNA work carried out in 2002 by scientists at Edinburgh University established that it was genetically distinct from the Narrow-lipped Helleborine, and therefore confirmed as *Epipactis dunensis*.

There are two forms of Dune Helleborine. One grows on the north-west coast of England and Wales on sandy soils and in the dunes, giving the species its English and Latin names. The other grows inland, most notably on the zinc-heavy soils of the Tyne Valley in Northumberland. For some years these plants were considered to be a separate Tyne Helleborine, but research published in 2002 confirmed it was the same species. Since then, populations have been discovered in strange places like gardens and university car parks.

We arrived in the Alyn Valley near Wrexham with the late-afternoon sun still hot on the tarmac. A small path passed through a stand of trees and out into a meadow. As we walked, I pointed out common bird's-foot trefoil and started teaching Suzie how to identify some of the commoner wild-flowers. She quickly picked up musk mallow, rough hawkbit and common centaury. I was in my element. We joked around and swapped stories about our families as we walked, content in each other's company.

I was showing her the various features of common knap-weed when she suddenly caught my eye and smiled: I paused, not quite recognising the emotion that had flitted so close. Feeling myself blush, I turned back to the knapweed in my hand and tried to gather my wandering thoughts.

After a short lesson about centauries, we entered the pine woodland, our feet slipping slightly on the sandy soil. The wood was stretched thin, light spilling in from above. We fell seamlessly into orchid-hunting mode, as if we'd been practising for years, and it immediately became an unspoken competition to find the first Dune Helleborine. We moved through the woodland in tandem, several metres apart, covering as wide an area as possible. Goldcrests seeped from the gilded treetops, occasionally visible as they flitted between branches. The air was warm and laced with thyme.

We slid down a bank, triggering miniature rivers of sand, and passed between willows and birches. Yellow bird's-nest sprung up like meerkats on the vantage point of small mossy hillocks. They are weird, beige-banana-coloured plants that are dependent on a fungus for food just like the Bird's-nest Orchid.

Suzie was the first to find a helleborine: not a Dune, but a

Green-flowered. It was tiny and had only three flowers. They hung down melancholically, a peaceful green in the low light. It is a similar species that is predominantly self-pollinated. The Green-flowered Helleborine is one of the wariest British orchids, confining itself to quiet, shady woodland rides not often visited by the botanist. It is so shy that in many cases the flowers never actually open: self-fertilisation occurs and the flower turns to fruit before the bud has burst open. Of the few plants in front of us, only one had decided to brave the outside world this year, and we left it to its own devices at the bottom of the slope.

Suzie and I continued our race to be the first to find a Dune Helleborine. In the end, this wasn't much of an achievement as the sandy woodland floor was, in places, carpeted with them. They were taller than the Green-flowered but slightly more yellow. The flowers, which weren't widely open, were small, dull and cup-shaped. Behind the pale curl of the landing pad, the dark inside of the lip glistened with nectar. In the intense sunlight that chequered the sparse woodland floor, the modest spikes blended into the sparse vegetation. Another of the shy late-summer helleborines, these Dunes were enjoying the peace and quiet in the depths of the woodland.

The next day, I woke early and continued north to Northumberland, leaving Suzie behind as she had to go to work. On the way, I stopped off in County Durham to visit the nature reserve in Bishop Middleham quarry. I arrived just as the sun started spilling in from the top of the steep, man-made cliffs.

Droplets of morning dew clung to each blade of grass so the ground glistened in the sun.

Abandoned in 1934, the magnesian limestone quarry rapidly became a haven for wildlife and is perhaps best known for its nationally important population of Dark Red Helleborines. These plants grow to three times the height of the less-sheltered colonies on the Great Orme. They were heavyweight helleborines: stunningly beautiful and well endowed with flowers. Each inflorescence was a mass of wine-coloured rubies and smelled like vanilla. Above the maroon lip, the golden-yellow anther caps acted as a warning or a sign of strength. In *Wild Flowers*, Sarah Raven describes them as 'the femme fatale of wild flowers: sultry, proud and exotically beautiful, but maybe not entirely friendly'. I wasn't sure about friendly, but they were certainly territorial. It was as if they were competing for the biggest hillocks; the tallest plants with the most flowers seemed to be found on the largest mounds.

A northern brown argus butterfly whirred silently past me and landed on some scabious. Next door were some Marsh Fragrant Orchids. At first sight, they looked just like the Heath and Chalk Fragrant Orchids: baby-pink and cylindrical, with horizontal sepals like the wings of a plane. But on closer inspection, I realised that there were tiny differences: a broad, open lip and a densely packed flower spike. The specific Latin name is *densiflora*, literally meaning 'dense-flowered'.

The quarry was a wonderful, botanical treasure chest, but I couldn't linger. I checked my watch and cursed. The timing for my next species was particularly tight.

294

For centuries, our native orchids have surprised, baffled and generally entertained botanists: populations grow and shrink, appear and disappear. But by the turn of the millennium, we assumed we'd found them all. In 2002, however, British orchidophiles were proven wrong. In the same study that finally put to rest the confusion surrounding the Dune Helleborine, scientists at Edinburgh University found that the population of helleborines on the Holy Island of Lindisfarne, Northumberland, were sufficiently genetically distinct from those on the mainland to be considered a separate species. Out of nowhere, they had discovered a species of orchid never before identified: Lindisfarne Helleborine, or *Epipactis sancta*. The holy helleborine. This newly ordained member of the family is one of Britain's few endemic species; the isolated population on Holy Island consists of approximately 300 plants. That's 300 plants on the entire planet, and all here on this small island. There are more giant pandas in the world than there are Lindisfarne Helleborines.

There is only one road onto Holy Island: a long causeway from the English mainland. Twice a day, when the tide sweeps in, this road gets covered by strong currents of choppy water, cutting the island off from the mainland and leaving tourists stranded for a few hours. Today's window: 10am – 4pm. I had to get on and off the island and find the helleborines before four or else get stuck there overnight and put my tight schedule for the next week out of sync.

With a recorded history that began with St Aidan in the sixth century AD, Holy Island has become a place of pilgrimage for Christians. Aidan, an Irish monk sent from the Scottish island of Iona, founded the Lindisfarne monastery. Northumberland's patron saint, Saint Cuthbert, was abbot of the monastery and later Bishop of Lindisfarne. The island is famous for its castle, the Lindisfarne Gospels and the long line of wooden poles between the mainland and the island that mark the final mile of the Pilgrim's Way.

Perhaps unsurprisingly, I had been to Holy Island with my family before. It had been a civilised three-sided tug-of-war between the beach, the dunes and the churches. At that time, though, I was equally interested in spending time appreciating the island's history as in scouring the dunes for plants. In the evening, when most of the tourists have left for the day, Holy Island is a place of serenity and peace that holds huge significance for my parents. After walking around the island, we would inevitably end up in the priory. At the time, I didn't entirely understand why they spent so long sitting among the ruins, but reasoned they probably didn't understand why I spent hours looking at orchids.

Two hours up the coast and I was driving onto Holy Island, a little later than I had hoped. The causeway was still slick and sandy but the waters had long since receded. The year I visited with my family, we had been too late to see the Lindisfarne Helleborines in flower, but I had still sought them out, managing to find several seeding spikes in the dunes. I therefore knew exactly where to find them.

It was a bright day and I was excited. The towering markers of the Pilgrim's Way stretched back towards the mainland, evenly spaced like telegraph poles in the sand. The wet sand was a blurry mirror of the sky above, punctuated sporadically by the familiar coiled castings left by lugworms.

I made the decision to go and find the orchids before eating lunch; there was a band of grey creeping over the mainland, so I wanted to make the most of the sun while it was still out. My stomach grumbled as I set off down a gravel track. A couple of buildings stood at the end: a small stone house and a tiny watchtower with a stunted turret.

The track soon gave way to sand underfoot and the tickle-scratch of marram grass against my legs. My decision to wear walking boots had been a bad one: not because of the heat, but because of pirri-pirri-bur, an incredibly well-adapted Australian native plant that is causing havoc as an invasive around the coasts of the UK. The dunes on Holy Island are rampant with it. After two minutes in the dunes, I glanced down and groaned: my shoes were covered in big seed burs that clung to anything and everything with their barbed hooks. Initially I fought it, stopping every ten metres to carefully pry the burs from my shoes and socks, but after a while I gave up, reluctantly deciding to start the world's biggest pirri-pirri-bur collection.

I circumnavigated a large dune slack and stopped for a while to take photos of some Marsh Helleborines. This was a species I'd seen the week before in Hampshire with my mother and Esther on a trip to an old waterworks. Miniature versions of the showy hybrids you find in hothouses, these orchids are a work of art. Inside their abundant flowers I saw pink washes of watercolour, deft strokes of yellow acrylics and inky dots and veins. The frilly, pure-white lip looks like a big Father Christmas beard.

I continued around the slack, confidently making my way over to the place where I had found the browning spikes of Lindisfarne Helleborine three years earlier. A dark green fritillary scudded over the sand in front of me, its wings a perfect horizontal plane burning orange and black.

After about ten minutes, I realised this wasn't going to be

as easy as I had thought. I walked back to the car park and retraced my steps, entering the dunes from a different direction, presuming I had simply gone wrong somewhere.

But I ended up in the same place. This was definitely where I saw them before, I was sure of it. As I set about combing the area between dune and slack again, it dawned on me, that the window of opportunity for seeing Lindisfarne Helleborines was becoming very small. The tide was already beginning to turn. By four o'clock, it would be impossible to drive back over the causeway to the mainland and to the campsite I had booked for the night. I really didn't want to sleep in the car. The next day I had to drive over to Cumbria; there would be no time to wait around until late morning for the tide to go out. The clock was ticking.

I had totally given up on lunch now and began pacing up and down, my eyes exploring the blanket of creeping willow for a yellowish helleborine. I texted Suzie and got a response which did little to help my nerves:

Imagine if you messed up on Lindisfarne Helleborine!!

It was two-thirty. I had one hour: it was the sand dune fiasco with Fen Orchids all over again. Other than a spot near the car park, which had already proved fruitless, I didn't know where else they grew.

The torment didn't last much longer. Relief swept over me as I caught sight of a tiny yellow helleborine among the willow. I knelt to examine the plant, just to make sure it was a Lindisfarne Helleborine and not a stray, unnoticed Dune Helleborine. Sure enough, the stem had a greenish-yellow base: no violet to be seen. The petals and sepals were various shades of apple green. It was unmistakeably a Lindisfarne Helleborine. One hour to spare.

I had walked past this plant two or three times, but that was hardly surprising as it was such a small one. There was no excuse, however, for the full-sized Lindisfarne Helleborine. How had I missed that?

I rushed back to the car, having collected the necessary photos and looked forlornly at my walking boots which were festooned with large brown burs. I had twenty minutes left on this beautiful island and I had to spend it removing these seeds. I ended up throwing my socks away.

That evening, I cooked my dinner on the beach at Bamburgh, looking out over the Farne Islands. The sandy wall of the dune provided a comfy backrest. In the distance, Lindisfarne Castle formed a grey landmark on the horizon. There were long lines of footsteps up and down the beach and small waves lapping at the shoreline as the tide advanced. Seaweed-encrusted rocks lay scattered along the sand. I realised they were slabs of concrete: an old road, perhaps, claimed by the sea. The sun set behind Bamburgh Castle, a jet silhouette against the flame-red sky.

The day after Holy Island, I raced back down the coast to Newcastle and then across to Cumbria to find Creeping Lady's-tresses. If I could find this little white orchid, I would really feel I was on the home straight. This was a crucial stage of the trip. If I failed here, it would mean a trip up to the vast pine forests of Scotland, something I didn't have time for if I was going to catch the remaining helleborines flowering in the south. I also needed to be in Ireland in a few days' time.

I had seen Creeping Lady's-tresses once before. My family and I had been on holiday in the Lake District and made a special trip to Penrith so I could seek out a thriving population bang in the middle of Whinfell Forest Centre Parcs. We had turned up, completely unannounced, and been confronted by security, who looked on with incredulity as my father asked to be granted entry so that his son could find some plants. Given we were quite obviously on holiday and there were three young children in the back of the car, the guards understandably took one look and refused my father's innocent request.

To my parents' credit, they persisted. After a lot of talking and various phone calls to members of the Centre Parcs team we were eventually let through to see the orchids, escorted by two members of staff.

A few days prior to my trip up north, I had learned that Centre Parcs now refused all access to the orchids unless people were visiting someone on site or had paid to stay there themselves. Clearly I hadn't been the only one snooping. This information came as a blow, because this is the only large population of Creeping Lady's-tresses in England.

Fortunately, I was handed a lifeline: there was a second, much smaller colony a few miles east in a small out-of-the-way nature reserve called Cliburn Moss. Jeff Hodgson, whom I had met while looking at Sword-leaved Helleborines in June, had provided me with some details. However, I quickly realised that this was a very risky strategy: Jeff had signed off his email with 'it will be very hit and miss I'm afraid Leif'.

I arrived in the late afternoon and pulled on my wellies

before walking into the woods. It was muggy and the humid air clung to me. The moist scent of crushed pine needles rose up with every step as I followed the soft, slightly springy path into the trees. A crow started in the canopy and I could hear the dripping of a spring or boggy flush somewhere nearby.

After a few minutes, I noticed a thinning in the trees to my right. Using Jeff's directions, I worked my way over to the end of the reserve, where I was instructed to leave the path. This was where it became a bit hit and miss, Jeff had said, and now I completely understood why: I was looking at a swamp. Pines and birches rose up at weird angles from the tangle of grasses, ferns and bilberry that made up the woodland floor. From the path, I could just about make out small dark areas where the grass hadn't taken over: pools of water.

Taking a deep breath, I left the path, walking slowly, cautiously, along fallen tree trunks, peat islands, anything that avoided the ominous dark water. I ducked under trees, jumped over stagnant ditches and more than once nearly ended up falling into duckweed-smothered pools. It was like walking into a scene from *Jurassic Park*.

After several minutes of navigating my way through the maze, I stopped and looked around me. The view was the same no matter which way I turned. There was no sign of where I was or how to get out.

Glancing around, I picked out a couple of stands of pine trees. This was where I would start searching, I decided. The Creeping Lady's-tresses would be on one of these islands where the pine roots had stabilised the peat, allowing the orchids to grow up from the thick carpet of needles that had collected over the years.

It was a task easier said than done, working my way from

one pine island to the next, uncertain of each step and sinking into wet peat, or worse, if I wasn't careful. It was quite beautiful really, when the sun came out and sent beams of light streaming between the trees. It became mysterious, idyllic almost. A rumble of thunder in the distance brought me back and I hastened my search. I didn't fancy being here in the middle of a summer storm.

After the third pine island, I was beginning to wonder whether I would ever find the orchids growing in this swamp. I was even starting to hatch ambitious plans to try and sneak into Centre Parcs. But on the fifth or sixth pine island, I struck gold with more than ten Creeping Lady's-tresses on the edge of a large pool of water. I sank to my knees, laughing with relief as I took in the short spindly spikes of small white flowers. Bizarrely, the flowers are densely hairy: an odd contrast with the delicacy of the petals. Like their arboreal masters, Creeping Lady's-tresses are evergreen plants and therefore often easier to locate during the winter when the surrounding vegetation has died back. However, they are only in flower for a short stint during high summer.

I was just seeing a fragment of the British population here. This small colony was merely an advance search party, scouting southwards. In the north of Scotland, Creeping Lady's-tresses form armies of thousands that conquer entire pine forests.

I spent the next hour attempting to photograph the plants. The fading light in the woods, not to mention their situation on the edge of the pool, made this an incredibly difficult task. The breeze that had been rustling through the trees before had picked up slightly, but the nearby bilberry was acting as a windbreak, allowing me some success with the tripod.

One hundred photos later, I took a break and began flicking through the camera to scrutinise my work. Not good enough.

I glanced to my left and sighed. There was only one thing for it. Suspending the camera around my neck I got up and walked over to the pool. It was covered in a layer of duckweed: miniature green lily pads. Below it, the water was jet black, not giving anything away. Carefully, not wanting to slip all the way in, I lowered my foot into it and tried to find the bottom. When I did there was a hiss and a cloud of bubbles raced to the surface. The water line was only a couple of centimetres from the top of my welly. I put my other leg in, so now I was standing in the pool, the icy water pressing in from all sides. My father would tell me I had to suffer for my art. I was now in the perfect position, with Creeping Lady's-tresses lining up in front of me behind veils of pond sedge.

I continued taking photos, trying to ignore the whine of mosquitoes, just occasionally standing up straight to slap one away. The sudden movement would send another stream of bubbles rushing to the surface and I would sink slightly lower, bringing the water perilously close to the top of my wellies. There was another rumble of thunder, closer now, but I ignored it: I was nearly done.

I was just about to start on the last plant, readjusting my footing in the pool and trying to ignore the numbness in my feet, when I suddenly became aware of a change in the atmosphere. The breeze dropped, as if a switch had been thrown. The air was dead still, cold now, and a menacing grey light had settled throughout the trees. Everything was silent. Even the mosquitoes had disappeared.

And then a crack of thunder split the sky above and well

and truly snapped me out of my daydream. I waded as fast as I dared back to the edge of the pool and climbed out, swinging my camera over my head and packing it roughly into my bag. I tried to remember which way I had come from. It all looked horribly similar. I set off briskly in what I guessed was the right direction. Where I had wobbled precariously before, now I was nimble as I walked deftly along dead tree branches acting as makeshift bridges.

Another thunderclap rent the sky. It was dark now and the air was dull and sticky. This was going to be a big one, I thought, very aware that I had left all my waterproofs in the car. I stopped and looked around me. There, to my left, were a few triangles of light that surely marked the edge of the wood. Brushing the first few spots of rain from my arms, I spotted the path and began making my way through the messy jumble of grasses that had woven itself into a web above the dark peat. Suddenly, I was back on the pine needle path.

I took one look back, taking in the dank swamp with its dark pools and hardy trees, suddenly lit up with a flash of lightning. It was with great relief that I turned and ran back to the car. I opened the door and threw myself in, moments before the heavens opened.

Five to go.

18

Ghost Hunting

Added to its few remaining sites will be the stanza
I compose about leaves like flakes of skin, a colour
Dithering between between pink and yellow, and then the root
That grows like coral among shadows and leaf–litter.
Just touching the petals bruises them into darkness.

Michael Longley, *The Ghost Orchid* (1995)

Buckinghamshire
July 2013

It began in 1854. Mrs Anderton Smith, the wife of the local vicar, set off along the Sapey Brook in Herefordshire to visit her brother-in-law. Edging down a small bank, muddy with recent rain, she caught sight of a tiny, colourless flower growing from the leaf litter in the deepest shade. It was very strange and so camouflaged that she kept losing sight of it. A single melancholic flower hung down, dull and creamy brown in the low light with just a hint of lilac. Unable to name it,

307

despite an extensive knowledge of the local flora, she leaned down and picked the first Ghost Orchid ever recorded in Britain. She enclosed the specimen in a letter to a local botanist, Mr H. C. Watson, by whom it was eventually named as *Epipogium aphyllum*.

They are small plants, a few centimetres in height, and often only offer up a single flower. The entire plant is a pale creamy brown, tinged with lilac, blending in perfectly with the deep carpets of dead leaf litter that it likes to grow in. The flowers are upside down, with the lip at the top, and are said to smell like bananas. Like the Bird's-nest Orchid, it doesn't produce chlorophyll and is entirely dependent on its fungal partner. It is almost ethereal in appearance: the epitome of inconspicuousness. From historical records, we know that it can flower at any point between April and October; that it seldom reappears in the same spot; and that it is usually destroyed by slugs. The flowers themselves, even if they do avoid the slugs, last only a couple of days. It was once known as the Spurred Coralroot, but the Ghost is more apt a name for this pallid lover of gloomy haunts.

Mrs Anderton Smith's discovery was exciting but then, as quickly as it had appeared, it vanished again. Twenty years passed and botanists began to wonder whether they had dreamed the whole thing. In 1876, another plant was found in woodland several miles away on the high road south-west of Ludlow. Francis Druce, an eminent twentieth-century botanist, tells us the story:

'Miss Lloyd found a specimen growing in the leafy debris which had been removed by a woodman from a trench in Ringwood Chase, Salop, to the adjacent bank. She took it to Miss Lewis who painted it, and submitted the drawing to Prof. Babington, who identified it as Epipogon [Ghost Orchid]; the specimen itself, owing to the carelessness of a servant, was destroyed. In 1878 the species was again found in the same wood, and I believe a specimen is in Professor Babington's Herbarium in Cambridge. For some years there was no further news, but in September 1892, I determined to make a search for the plant. My first intention was to explore the Tedstone Delamere locality, but accidentally meeting Mr. J. Gilbert Baker at the Oxford Station I went with him part of the way and then went on to Ludlow. There Miss Lloyd very kindly accompanied me to the wood where she first found it, but a close search failed to discover any trace of it. The next day, the third of September, a renewed and arduous search was made, when at length, in quite a different part of the wood, well within the Shropshire boundary, in dense shade of oaks on a level part of the steep, sloping wood facing northwards, a small, solitary flowering example was found which I had the pleasure of showing to Miss Lewis, and it is now one of the chief treasures of my Herbarium. Although many botanists, especially the lynx-eyed Mr. Wedgewood, have repeatedly searched the wood, no other example has been found, and during the war the portion of woodland where the plant grew was cleared of timber.'

The hunt went on. In 1882, Mary Lewis wrote in a letter to Babington: 'owing to an outbreak of typhoid fever in Ludlow I was unable to have a good hunt for the *Epipogium* [Ghost Orchid] last August. Neither of the two finders have a dried specimen. Miss Lloyd's was thrown away by mistake.

Miss Peele tried to get it to grow in "their" garden – of course unsuccessfully. It had been seen by a retired chemist named Cocking living here but he always thought it an abortive Bee orchid saying he never examined it as the colour etc. was so poor.'

Twenty years later, Druce himself discovered another. But it didn't persist. The Ghost Orchid was rapidly gaining a reputation for these rare appearances.

After the First World War, the hunt for the Ghost Orchid gained momentum. People were desperate to become the next person to find one. In 1924, botanist and plant hunter George Claridge Druce was on holiday in Jersey when he got a call from his friend Herbert Smith, who breathlessly described the two spikes of *Epipogium* standing in a vase in front of him. They were found and collected by a local school girl – not in Shropshire, but in Lambridge Wood, just west of Henley-on-Thames in Oxfordshire. Druce couldn't believe it. Without a moment's hesitation, he abandoned his holiday to head to Henley, where he met the girl, a Miss Butler, who took him to the wood. He spent hours combing the site and after days of searching found two spikes.

A couple of years later, botanists Eleanor Vachell and Francis Druce visited Miss Butler, who wouldn't relinquish the specimen she had sitting in a jar on the window-sill. However, she agreed to take them to the place she had found the plants, but they couldn't find any flowering spikes. Vachell had written at the time of searching that 'excitement knew no bounds'.

In 1931, four miles away from Lambridge Wood, another school girl, Miss Vera Smith (later Vera Paul), discovered a Ghost Orchid in a new locality: Great Bottom Wood. This was widely publicised and she recounts the original discovery: 'On June 30th 1931, my father and I were looking for Fly and Butterfly Orchids when I saw an orchid which I did not recognise growing out of the middle of an old tree stump. Not realising how rare my find was, I picked it. Dr. Somerville Hastings was at home, and we discovered the name and status of the plant. Our excitement was so great that we literally wrapped the plant in cotton wool and took it to the Reading Museum, where it was photographed and preserved. It was 9½ inches high.'

In 1963, she tells us about another visit:

'It was almost dusk when I arrived. Imagine my surprise when I saw not one, but five, spikes, standing in a row near the original stump where the plant first flowered. Unfortunately, the slugs had already been busy, and three of the spikes had been eaten through at the base, so that they were standing by the support of the beech leaves among which they were growing. The parts of stems and one flowering shoot were gathered up and are now in the possession of Reading University. One unusual feature of the plants this year was the appearance of the underground stolons above the surface of the ground. Further examination of the area showed that the plants extended along several yards of thick humus and one other small group of flowering shoots were coming out from underneath a stone.'

In 1953, there was a shift in focus as Ghost Orchids started appearing in Buckinghamshire. Rex Graham found twenty-five flowering spikes in one season, the first spotted over the bowl of his pipe one afternoon, in the Chiltern beech woods. This continued to be a productive Ghost site for thirty-four years.

In 1987, the Ghost Orchid disappeared, seemingly for good. Over the next twenty years there were various unconfirmed sightings, but none were proven. In 2005, eighteen years since the last record, the Ghost Orchid was declared extinct. The charity Plantlife even focussed their new campaign for saving Britain's wildflowers on the loss of the Ghost. But then, in 2009, it appeared again. On the twentieth of September plant hunter and motorbike company owner Mark Jannink discovered a single spike in the woods in Herefordshire. Upon spotting the plant after weeks of searching, he was said to have exclaimed, 'hello you – so there you are!' This was the last time the Ghost Orchid was seen in Britain to date.

In the UK, the Ghost is extremely high maintenance. It is said to flower following a wet spring and in soils that are damp from April onwards until the end of October. The canopy must be nearly complete in order to reduce competition from photosynthetic plants. To make things even harder for us, the Ghost is thought to be able to flower underground, buried in the humus created from the leaf litter. A plant that can flower underground? No wonder it's so rarely found.

Despite making the decision to exclude the Ghost from my list, I certainly wasn't going to let it go unsearched for. With July fading into August, Suzie and I decided it was time to start hunting for my bonus fifty-third species.

As we entered the wood, Suzie and I were immediately swallowed up by the gloom. Our eyes adjusted to the low light, revealing a cathedral of beech trees. The woodland floor was a bare sweep of ochres and beige, decades of dead beech leaves strewn everywhere. There was no green vegetation. Thirty metres above us, sunlight was seeping through the thin canopy, creating an eerie glow that settled throughout the wood.

This was the Buckinghamshire site where Rex Graham found twenty-five flowering spikes in a single season, way back in 1953. I figured it was worth checking out. Perhaps I could repeat the fortune I'd had with the Lady's Slipper in Yorkshire. Following directions from Sean Cole, a true orchid-ophile and Ghost-obsessive, we began searching in earnest. Surely it would make an appearance for me during my summer's quest?

Either side of the tarmac the ground fell away almost immediately into ditches. Fallen branches had been left to rot. Mosses had claimed the damp banks, collecting the water that trickled slowly down the slopes.

We left the road, splitting up to cover a greater area. While Suzie pottered further into the wood, I cautiously picked my way along the ditch, stopping every couple of metres to bend down and scan the undulating woodland floor. The crispy leaves scrunched underfoot, wafers of a summer long gone. It had been very dry recently: conditions weren't ideal. Was the Ghost here? I got the feeling that this was an activity that required silence. The Ghost was skittish. Too much commotion and it would melt into the leaf litter, never to be seen again.

There was an unnerving silence in the wood as noise from the outside world was muffled by beech. I'd momentarily lost sight of Suzie and soon felt like I was being watched. It had been a long time since I had been scared and alone in a forest. A twig cracked somewhere behind me and I surprised myself by wheeling round. I was imagining things. This place was getting to me.

Twenty metres away, Suzie pulled a hand torch from her bag and, holding it just above the ground, began sweeping the beam in an arc. I teased her for doing this, but I'd read it was a neat trick that made spotting Ghost Orchids easier. They would shine pale in the bright light.

Halfway down the hill, I found a council of Narrow-lipped Helleborines. Green and spindly, much like the Dune and Lindisfarne Helleborines I'd seen over the previous few days, the lip of this orchid doesn't tuck underneath the flower, but is unfurled, tapering to a fine point. It is fairly rare, and resolutely refuses to grow anywhere other than in southern beech woods.

Suzie moved unhurriedly down the hill, head cocked to one side as she crouched down to survey the brown sea of leaves. Every now and then she would sit cross-legged on the forest floor and rest her chin on her hands, watching and waiting. She was conspicuous with her red rain jacket, pink polka-dot wellies and blue-and-white umbrella: the Ghosts would see her coming a mile off. Her hair was tucked into a ginger bun, a few loose strands falling by her face. She glanced up and caught me grinning at her before continuing her search.

Ghost hunting was unlike anything I'd attempted that year. It was a completely different way of searching for orchids. You are balancing the fact that you are searching for something that probably isn't there with the thought that you just could find a Ghost Orchid at any second. Every time I crouched down, I was gripped by a fresh sense of excitement. Every year, a small group of dedicated Ghost hunters spend hours on end combing damp, dark woodland across the Midlands, living with the hope, a small hope, of meeting the Ghost Orchid. Finding one makes them a kind of celebrity within the botanical community, but they aren't after fame. They're driven by the need to find something so fragile and rare that discovering one is less than a once-in-a-lifetime occasion.

After three hours of Ghost hunting, we gave up and began wandering back to the car. On a few occasions, I had glimpsed a tiny spike emerging from the leaf litter and my stomach had dropped, but they were only ever small mushrooms or twigs. I kept imagining the moment over and over in my head. If I found a Ghost, what would I do? How would I react?

As we were nearing the car, we came across two Broad-leaved Helleborines growing right next to the road. Someone had positioned a fallen branch in front of them to prevent them being run over. These helleborines have the same basic flower shape as their cousins, but its lip is curled under the flower unlike the Narrow-lipped Helleborines I had found earlier. The Broad-leaved Helleborine is one of the commonest orchids in the country and is highly variable. You'll find tall robust plants, thin weedy ones, some with pallid green flowers, rich purplish flowers or usually a mix of both. While most often associated with woodland, they can be found in a wide variety of locations and are a highly adaptable species. Here they were in heavy shade, but in the past I had also seen them completely out in the open. Most orchids are exasperatingly picky about where they grow, but the Broad-leaved

Helleborine takes what it has and gets on with things. I admired that in this humble-looking orchid.

I had to admit there was a certain magic about the wood and its enigmatic resident who may or may not be here. I could see how the longer you searched, the more you had to search; it was addictive, and I understood how people became so obsessed with this tiny, spectral plant. Every moment could be the moment.

Since its discovery in 1854, the Ghost has tormented us and flirted with our imaginations. Here's hoping the dedicated team of volunteers monitoring sites throughout the summer can rediscover this enigmatic orchid.

As we got back to the car, my imagination caught up with me and the feeling of unease I had been trying to ignore swept over me. I no longer wanted to be anywhere near this place. It was so empty. As we left, I couldn't shake the feeling of being watched and my spine was tingling as we made for the sunlight.

Spooky.

19

August Orchids

'And what do you think of this for an Orchid? Thought the
Orchids were all over, did you? This is one of the later of
them – the Purple Helleborine'

Edward Step, *Wild Flowers in their*
Natural Haunts (1899)

County Mayo, County Leitrim and Wiltshire
August 2013

Rain was lashing the windows. My journey up the M6 was
slow, a blurry train of red tail lights and the drumming of rain
on the roof. I turned the radio on and half-listened to the reel
of pop music, news bulletins and over-enthusiastic adverts. I
had spent too much time in recent weeks on this motorway,
listening to Radio One and sliding along in bumper-to-tail
traffic jams. I was, in a way, glad it was nearly over.

Suzie and I were going to Ireland to hunt down the only
species I had now never seen before: the Irish Lady's-tresses.

I pulled off the motorway and rolled into a service station car park, our agreed meeting point. I was ten minutes early. The rain continued to hammer against the windscreen, as if desperate to break through. It was utterly bleak. If the first day of August was like this here, I was concerned about how wet it would be in Ireland.

Outside, families were dodging puddles in the car park. It was packed full of estates and people carriers piled high with roof boxes and bikes. People were making their way north: the Lake District, perhaps, or Scotland. Overly enthusiastic parents were doing their best to cheer up bedraggled children. Outside the entrance to the service station, a sign advertised half-price folding beach chairs and buy-one-get-one-free on a range of sun creams. The epitome of the British summer, I thought.

A knock on the window startled me out of my daydream. It was Suzie. Behind her stood a bespectacled man in a dark coat. He was well-dressed and had water dripping from a greying beard: her father, presumably. I jumped out and greeted them enthusiastically, once again painfully aware of how posh I sounded. He seemed surprised when I shook his hand. I gave Suzie an awkward hug. I suddenly realised how much I'd been looking forward to seeing her. It had only been a week.

Having loaded her bags into the car, we bid her father goodbye and rejoined the motorway, heading west across the top of Wales towards Anglesey. The rain receded slowly and by the time we reached Holyhead it was dry, though the sky remained bullish and angry. The ferry terminal was quiet, and Suzie entertained herself by reading my travelogue out loud. Her laugh permeated through my moody embarrassment and we both ended up giggling about my turns of phrase. We had planned a whole week together to maximise our chances of finding Irish Lady's-tresses. I wouldn't be able to come back again.

We sat outside in the freezing sea air, watching shearwaters skim over the waves and gannets fishing for their evening meal. Suzie was scanning the horizon with her binoculars, her hair swirling about in the wind. A tern bombed past and she grabbed my arm and pointed. She was in her element.

The next day, after overnighting in Dublin, we arrived in a little village called Boyle in County Roscommon, where I had booked us a B&B for the night. We had arrived earlier than expected, so decided to go and have a look for Irish Lady's-tresses over the border in County Mayo.

The journey over to Lough Cullin was a repeat of the previous day: the driving rain was so heavy that it became near impossible to see out of the windscreen. Two pinpricks of red light were all I could make out of the car in front. I couldn't imagine finding any orchids in this weather.

We passed through Foxford and followed a road that ran between Lough Cullin to the south and the much larger Lough Conn to the north. Even bothering to stop on the northern shore seemed like a pointless exercise: the rain was intense.

These two lakes are known for their salmon and trout fishing and are connected to the Atlantic via the River Moy. According to Gaelic mythology, the two loughs were formed when Fionn mac Cumhaill, a mythical warrior-chief, was out hunting with his hounds, Conn and Cullin. Upon encountering a wild boar, the two hounds gave chase across the Irish wilderness. As it fled, water poured from the feet of the boar and, over the course of a few days, it formed a large lake that claimed the life of Conn. Cullin took up the hunt, but suffered the same fate slightly further to the south.

As we sat in the car park, the rain began to abate. Slowly at first, but then it suddenly stopped altogether. Now that we could see, we took in the beautiful view of the lough and surrounding vegetation: boggy pasture and damp grassland, perfect territory for Irish Lady's-tresses.

The glassy surface of the lough stretched far into the distance, rutted with dark rock. A gull floated lazily from right to left and disappeared behind a thin green spit that tailed into the water. Between the shore and the woods was a swampy no-man's-land littered with rugged boulders. The ground was boggy and water-logged; miniature streams trickled quietly from lush vegetation to bare sand and the lough beyond. Young saplings were taking their first tentative steps away from the forest.

Minutes after the rain had stopped, the clouds burst open, unable to contain the streams of evening sunlight that flooded the scene in front of us. We took the bait. Leaping from the car, we pulled our wellies on and crunched down the track to the shore, tiny waves lapping soundlessly at our feet, and started following the water's edge east. The air was cool and smelled of fresh rain.

The plants here were incredible: a wonderful assortment of knotted pearlwort and marsh cinquefoil, purple loosestrife and bristle club-rush. Sneezewort, which grew everywhere, provided an opportunity for some extreme botanical geekiness and I jumped on it. Suzie listened attentively as I told her how it produces an essential oil used in herbal medicine;

that its leaves are used for insect repellent; and that, like many other plants, the flowers in the inflorescence are arranged in a Fibonacci sequence to maximise the number it can pack in. Indeed, the inflorescences of sneezewort, and all species in the daisy family, actually have two types of flower. Consider an ordinary daisy that you might find in your lawn. When playing 'she-loves-me-she-loves-me-not', each white petal you pick off comes from a separate flower. Together with the yellow flowers bunched up in the middle of the daisy, they form an inflorescence.

I glanced at Suzie after I had told her all this, afraid that I had overstepped a mark and freaked her out a little. I had never divulged such embarrassing and nerdy information to anyone before, let alone to a girl. But instead of labelling me a loser, she seemed impressed. On a whim, I reached out and clumsily grasped her hand. I flushed crimson, stunned at my spontaneity and forwardness. A look of surprise flickered across her face, but only for a moment, and we continued walking, hand in hand, both feeling too awkward to acknowledge what had just happened.

Fortunately, just as I was about to burst with the tension, I spotted the first Irish Lady's-tresses. I yelped, both from relief and excitement. Releasing her hand, I raced over to them and knelt down in the silt. They are majestic orchids: pale-green stems are topped with pure-white flowers twisted three ways into an upward-reaching spiral. Each one bore a broad white tongue. They were tall and surprisingly robust, yet closer inspection of the flowers revealed a delicacy normally associated with intricate wedding cake decorations.

I couldn't stop grinning.

While Suzie had her turn taking photos, I kept walking,

searching for more and carefully identifying plant after plant and putting together an impressive species list. One of the most magical plants in this habitat is grass-of-parnassus, a stunning member of the family Celastraceae, a close relative of the saxifrages. It isn't anything like a grass. Easily recognisable from its single-stemmed white flowers, it has five petals, each patterned with delicate, silvery channels. It gets its name not from its appearance, though, but from ancient Greece where the cows of Mount Parnassus took a liking to it.

I found another six Irish Lady's-tresses, but none as impressive as the first. Some spikes were jagged combs, uneven and pointy. Others were small with only a few white-tongued flowers. Yet they all rose with splendour from the tangle of pennywort, alder and mint.

I looked back to see Suzie crouched down and trying to position a clear plastic umbrella to aid her photography in the wind. Her tripod and camera stood off to one side, seemingly untouched. I teased her about being too keen, to which she replied with a torrent of insults. I smirked back, defeated.

The sun was glinting at us from the edge of a cloud now and the sky's reflection was painted across the expansive canvas of wet sand under our feet. On the horizon, fresh rain clouds were piling in: a dramatic rolling grey front. Cetacean-like rocks rose from the flat surface of the water, black against the sun's bright glare. Lough Cullin was hauntingly beautiful, tucked away in this remote corner of Ireland.

Incredibly, I had now reached fifty for the summer. Fifty! This was a huge milestone, as over the course of my teenage years I had now seen every native orchid species in Britain and Ireland bar the elusive Ghost. It was certainly a cause for celebration. Both Suzie and I were quietly satisfied with finding Irish Lady's-tresses all by ourselves, without needing the guidance of a pointing finger.

That pointing finger was lined up for the next day. After leaving Lough Cullin, the weather had immediately turned again and the rain remained torrential until the following morning. We had driven east to a small town called Drumshanbo in County Leitrim, which takes its name from the Gaelic for 'ridge of the old huts'. Away from the hustle and bustle of tourist Ireland, Drumshanbo is a proper Irish market town, situated at the southern tip of Lough Allen: a colourful corner of Leitrim surrounded by undulating hills and luscious green woodland.

David and Frances Farrell have been running a voluntary conservation project around Lough Allen for the past decade, recording everything from Daubenton's bats to red-breasted mergansers and Small White Orchids. I had contacted David to see if we could arrange a day to go looking for the Irish Lady's-tresses, to which he had enthusiastically agreed.

The Farrells' house was lost in the rolling Irish landscape. It was small, red-brick and sprawling. Each room appeared to have been bolted on to the last, the result being a rambling, slightly run-down maze of a property that looked to have seen better days. We decided it was wonderful.

As we slowed to a halt at the front of the house, David and Frances came out to greet us. I climbed out and found myself immediately shaking hands with a very enthusiastic David. Their dog was crowding around our feet and I glanced over to see Suzie in the embrace of Frances. The Irish were certainly friendly. We were welcomed into their home like old friends and ushered through to the kitchen where a kettle was whistling away on the stove. Pots and pans teetered in tall stacks on the red-tiled floor. Taking our seats at the table, Frances set two steaming mugs of hot tea in front of us and sat opposite, brimming with questions.

They were fascinated by my summer's quest and honoured that I had chosen to come here to see the Irish Lady's-tresses. These orchids were their personal pet project.

Over the last decade, they have religiously surveyed the lough shore for Irish Lady's-tresses, discovering new populations and counting individuals throughout the flowering season. I admired their enthusiasm and drive to help this rare and declining orchid, bearing in mind all the other wildlife that the lough has to offer. Combining their knowledge with that of experts, they have slowly built up the pattern of distribution around the shore and been rewarded with multiple new sites. Over the years, they have also developed their understanding of the threats to its survival, working with local land owners to control grazing and therefore avoid damage to the plants in the flowering season. As the orchids are active above ground all year round, though, it turns out natural flooding offers the best protection against overgrazing. The Farrells' project has provided invaluable insights into the dynamics of this species in the quest to protect it from extinction in Ireland.

When we finally got there, Lough Allen didn't disappoint. It was a wide stretch of water punctuated by small tree-covered islands. The grassy shores were damp and springy underfoot, banks of reeds swarming out of the water. This, David told us proudly, was prime tresses habitat.

Suzie and I followed David down the hill, trying not to laugh at the enormous hole we'd noticed in the side of his shirt, while Frances went on ahead to check some new sites on the shore. We passed sneezewort in the damper ground and devil's-bit scabious on the drier mounds. Reaching the shore, we turned south, following the high tide line. After five minutes, we found the first of fourteen Irish Lady's-tresses, dotted along the shore in the long grass. Just like the plants at Lough Cullin, they ranged in size and robustness.

The meaning of the Latin name *Spiranthes*, with which David and Frances fondly referred to their treasured plants, is an obvious reference to the spiralling inflorescence, the way in which the three ranks of flowers twist skywards. Ted Lousley describes its 'resemblance to old-fashioned ways in which ladies once twisted their hair' – hence the name 'lady's-tresses'.

The first Irish record was made in 1810 but wasn't published until 1828 by J. E. Smith in his *English Flora* – a bizarre inclusion given it had been found in County Cork. For the next hundred years, it remained an Irish orchid, until in the early 1920s it was found in Scotland on the Isle of Coll. In 1930, it turned up on Colonsay and, one by one, other Hebridean islands acquired small satellite populations. Now Barra, Mull and South Uist all boast Irish Lady's-tresses. They're more genetically diverse than the Irish plants too. But perhaps the biggest surprise came when it was discovered growing in Devon. Smith's decision to include it in his English flora was vindicated, after a routine botanical survey on Dartmoor turned up seven plants in 1957. The population persisted for several years, but has not been seen there since the early 1990s and is presumed extinct.

Irish Lady's-tresses cannot be found in Europe outside of Britain and Ireland, but occur across North America. Synonyms include Hooded, American or Drooping Lady's-tresses. It was first found in Alaska while the region was still a Russian territory and its species name, *romanzoffiana*, commemorates Nicholas Romanzoff, a former minister of state for Russia who was a benefactor of science. So how did it get here? Is it a relic from pre-glacial times? Or was it carried across the Atlantic more recently by migrating geese? The latter seems fanciful, and the most widely accepted theory is that it has grown here since before the last ice age. However it got here, Britain and Ireland have the only populations this side of the Atlantic so its conservation priority is high.

However this orchid is a bit of a nightmare for conservationists. While the number of sites is increasing, largely due to more effort being put in to find them, overall numbers seem to be falling. It is given to vanishing unexpectedly and making fleeting appearances here and there. Nor can it make up its mind about which grazing regime suits it best, or which weather conditions. Grazing regimes are very difficult to perfect: too much and the orchids will suffer, too little and they will be out-competed by tough swards of grass. None of this is very helpful for eager conservationists trying to halt its rapid and sudden decline.

Just as we were about to turn back, we found an enormous spike supporting an inflorescence of at least thirty flowers. Its whipped peaks of snowy white spirals were near-perfect mimics of the mollusc shells scattered on the lough shore. It was blustery and Suzie's clear plastic umbrella was very useful as we struggled to capture this beautiful white bloom. The reed beds to the south were shimmering like a field of barley. Gulls circled overhead, their cries snatched away by the wind. The icy water chopped and slapped at the shoreline.

Meeting David and Frances and hearing about their ongoing commitment to this orchid was inspiring. This was a small bubble in the middle of rural Ireland, and one that was in danger of being burst. Frances explained to us that water quality has been a growing concern for them: contamination is resulting in an increasing frequency of algal blooms in stagnant backwaters. The intensification of land use around the lough is also cause for concern, as valuable tresses habitat is being lost.

But despite these threats, the Farrells were full of hope and extremely encouraged by our visit and interest in the plants there. It was almost as though our presence was confirming that this was a worthwhile project and not simply a futile task done purely for personal gain. I certainly made no secret of how exciting it was to be there. I was, and remain, extremely grateful for their hospitality and eagerness to help us out. Most of the land around Lough Allen is private so we wouldn't have had the opportunity to see the orchids there were it not for the Farrells.

The Irish Lady's-tresses is holding onto its small territory in the north-western corner of Europe, but the fact we know so little about its ecology and distribution makes it problematic. With climate change, plant species might begin migrating north, but already at its limits, there isn't anywhere for this orchid to go.

We must also learn from the past. A close relative of Irish Lady's-tresses is Summer Lady's-tresses, the only orchid to have become extinct in Britain. Its delicate tress is slighter than *romanzoffiana* and it grows in peat bogs, thriving on high moisture levels like its Irish cousin. Last seen in the New Forest in 1952, it is a constant reminder of the dangers of collecting and habitat destruction. It was once quite abundant in select areas; E. D. Marquand noted that in 1901 he 'saw half an acre of bog perfectly white with these flowers'. But Summer Lady's-tresses began its decline in the twentieth century – land drainage and over-collecting were the nails in the coffin. However, it survives on the shores of northern France, so it is strange that it hasn't successfully re-established itself in Britain; we might expect it to turn up again as climate change progresses. However, due to the unfortunate propensity of a few botanists to illegally reintroduce orchids, any natural occurrence in its former haunts will be looked upon with suspicion.

The rest of the week came and went fairly leisurely. Now with a fortnight or so until the Violet Helleborines were due to flower, I had a bit of time to relax. Suzie and I drove north and explored Northern Ireland, paying visits to Portrush and the Giant's Causeway. We visited another beautiful site for orchids on the shores of Lough Neagh, the largest inland body of water in the British Isles, to the west of Belfast. As legend has it, a fight between two giants – one Irish and one English – broke out across the Irish Sea. The Irish giant picked up a large rock and threw it at the English giant, but it fell short and crashed down into the sea. It's still there, and now known as the Isle of Man. Lough Neagh fills the hole from which the Irish giant took the rock. The Irish Lady's-tresses here were few and far between and more difficult to pick out as the mist began to descend on us. It made us wonder how this orchid could survive in this cool, rainy place where so few insects dared to brave the weather.

Over the next few days, we went for walks along the cliffs and watched the bird colonies, keeping an optimistic eye out for whales on the horizon. I taught Suzie some more botany and despite her best efforts, I beat her at mini-golf. It was nice to be able to blow off steam and enjoy a few days hanging out, rather than rushing around the country and worrying about whether or not I would catch the next orchid in time. I had forgotten what life was like when I didn't have to think about orchids. I had become so wrapped up in my trip, so obsessed with these fifty-two species, that I hadn't really thought about what came after.

One evening, we decided to drive to the cliffs and cook dinner while watching the sunset. Suzie is one of the fussiest eaters I have ever met. She would remove the membrane between each layer of an onion before cooking it, and the tiny seeds from the surface of a strawberry. We cooked sausages on my camp stove, hungrily devouring them as the sun began to dip towards the horizon. I took her hand and led her to the fence where we could get the best view. And with the same spontaneity that had so surprised me at Lough Cullin, I kissed her.

We returned to England midweek and I began filling the time with short trips from home to visit local helleborine populations. I went to find more Green-flowered Helleborines in Salisbury to complement the three we had found in the Alyn Valley. I travelled to Dean Hill near Whiteparish to hunt down more Broad-leaved Helleborines, and found a wonderful array of colours: ruby red, sandy yellow, rosy pink and lettuce-leaf green. Incredibly, they all grew within a couple of metres of the busy A36. This was probably as dangerous as my orchid hunting would get.

Suzie and I returned to the Chilterns for more Ghost hunting. We decided to visit Great Bottom Wood, where Vera Paul made her famous discovery in 1931. The sense of competition was electric. It would be desperately unfair if she found a Ghost before me. I refused to think about her taunting me if she did: it would be unbearable. We searched for hours, but other than a suspicious-looking hole, found nothing.

After seeing the Irish Lady's-tresses, the remainder of my quest seemed easy. Finding that one so effortlessly had surprised me: maybe it was practice after a long three months. As August wore on, my thoughts returned to orchids and settled on the penultimate species of the season: the Violet Helleborine.

With my adventure drawing to a close, I was keen to get my whole family along to see this woodland guardian. I had visited Savernake Forest in north Wiltshire once before, trekking through on my Bronze Duke of Edinburgh expedition. That had been in June, though: far too early for this late-summer orchid.

Away from the congestion and tail-gating of Marlborough, Savernake is ancient, wild Wiltshire. After the Norman conquest, the forest was put into the hands of Richard Esturmy and has been passed down, from parent to child, in a thirty-one-generation line of hereditary wardens. It has remained within the same family for over a thousand years and is Britain's only remaining privately owned forest. Once amounting to 4500 acres, Savernake was visited regularly by King Henry VIII, who enjoyed deer hunting in the forest. Sir John Seymour, the warden of the wood at the time, would later become the King's father-in-law as his daughter Jane was crowned queen after the execution of Anne Boleyn in 1536.

A royal forest steeped in history: where better to go in search of the Violet Helleborine? I gathered my family up and we all squashed into my Ford Focus for the drive. The high summer sun was baking the tarmac so it was a relief to slide into the dappled shade of the forest in the early afternoon. The brambles hung low, heavy with tight green fruit, and a blackbird fled down the lane in front of us. I hadn't even stopped the car before I saw my fifty-first orchid of the year. By the side of the road, two metres across the parched leaf-littered soil, stood a pale, glow-in-the-dark Violet Helleborine.

I pulled the car off the road and jumped out, ready to photograph this, perhaps our most mysterious helleborine. My sisters came over to take a look but lost interest within a couple of minutes and wandered off to explore the forest. But I was transfixed. Violet Helleborines rose up all along this short stretch of road running through the trees, their stems flushed indigo. They are mysterious orchids, preferring to grow in dark, heavily shaded woodland where you are unlikely to find many other herbaceous plants. It was in many ways reminiscent of the other helleborines I had found, but the dark-violet stems and pale green-yellow flowers were unmistakeable. Its flowers are ghosts, glowing eerily in the dim light as they bend round tree trunks, trying to catch a glimpse of who or what has invaded their quiet sanctuary.

First discovered by the Reverend Dr Abbot in Worcestershire in 1807, numbers of the Violet Helleborine diminished over the following century with the decline of ancient woodland. Unlike most orchids, though, this one is thought to have benefited from the abandonment of coppicing in the twentieth century. As the trees grew, and the canopy developed, the Violet Helleborine stole back in.

My family slowly meandered back to the car where they set up my camp stove and began boiling the kettle for a cup of tea. They left me to my photography, and I spent an hour in this tiny corner of Savernake, enjoying the tall spikes of Violet Helleborine as they swayed in the light breeze. At the end of the road, I saw streams of walkers and cyclists. No one took any notice of the orchids lurking by the roadside. But that was exactly how the Violet Helleborines wanted it.

August ended somewhat painfully. I had finally worked up the courage to ask Suzie out. I wanted her to be my girlfriend more than anything. More so, even, than completing my list of orchids. She was a beautiful, like-minded, hilariously entertaining orchid-hunting companion.

One final Ghost-hunting expedition at the end of August had brought us back to Buckinghamshire. We found more Violet Helleborines, again by the side of a road. One plant, nicknamed the 'Big Daddy' by local enthusiasts, had a phenomenal thirty flowering spikes. Cole told me that a few years previously it produced fifty-two spikes.

We searched long and hard, ten metres between us as we swept the wood in a start-stop stuttering motion. Suzie was looking intently at the ground while I shook with nerves as I built up to asking her out. I had tried and failed to breach the topic on our final night in Ireland. I had been too faltering: there was too much riding on it and, certain she would say no, I didn't want our new-found friendship to end. But it was time. I had to ask. It was nearly the end of the orchid season, and soon there wouldn't be an excuse to meet up any more.

It all happened very quickly, but hearing her swift rejection was just as painful as I had expected. We wanted different things, she said. We were living halfway across the country. I was going to university. All logical arguments, but they only made it more difficult to take. Perhaps I had known all along that it wouldn't happen.

We walked back to the car in awkward silence. I felt like I had failed. Perhaps, somewhere deep down, I had hoped that I would get something more than orchids out of this year. My yearning for a relationship, one that could truly accommodate my love of botany, had manifested in feelings for Suzie. It was inevitable really, though I genuinely hadn't foreseen this possibility at the start.

I dropped her at Banbury station and saw her onto the train. She waved goodbye, and I knew that this was probably the last time I would ever see her. It had been a fleeting, magical couple of months, but that was all it was.

20

Spirals by the Sea

'We shall not cease from exploration,
And the end of all our exploring
Will be to arrive where we started
And know the place for the first time.'

T. S. Eliot (1888–1965)

Dorset
2nd September

I needed a pick-me-up. It was the end of a long summer of travelling; running around and trying to tick everything off a list. Suzie's rejection was still raw. My life for the last five months – for the last twelve years – had been defined, directed, utterly dictated by orchids. They'd had total control over me for so long, governing my every move.

I'd had a successful summer, but there was one orchid I still had to find: the Autumn Lady's-tresses. Edward Step, in *Wild Flowers in their Natural Haunts*, introduces it as 'the last of the [Orchid] tribe we shall meet this year'. I liked the idea I was going to 'meet' them.

After the excitement of seeing Violet Helleborines and the realisation that I was only one orchid away from finding all fifty-two, there was a long wait. August slipped into September with a worrying lack of Autumn Lady's-tresses. I was aware that the late season might still be affecting things, but, for the most part, the orchids had more or less caught up with themselves. The Irish Lady's-tresses and Violet Helleborines had flowered when I would have expected them to in a normal season and so during the final week of August, when hundreds of Autumn Lady's-tresses would normally be coming into flower, I was a little surprised that I still hadn't heard anything.

The schools went back at the beginning of September and, for the second time in as many years, I was left feeling rather lost. The summer was drawing to a close: blackberries were ripening into plump purple clusters in the hedgerows, the hay meadows on the hill were yellowing, and the combine harvesters were busy reaping the rewards of a warm and bountiful growing season.

Autumn Lady's-tresses is one of the commoner orchid species in the UK and so seeing it wouldn't be too big a problem once it eventually arrived on the scene. After much deliberation, I decided to go back to Durlston Country Park to see number fifty-two. It seemed satisfyingly symmetrical to try and end my adventure in the same place that I had started it five months previously.

Autumn Lady's-tresses was the first orchid, along with the Common Twayblade, ever to be recorded in Britain, as early as 1548. William Turner wrote in his *Names of Herbes* that it 'groweth beside Syon [Sion House, opposite Kew Gardens] …it bryngeth forth whyte floures in the ende of harveste, and it is called Lady traces'. Its fondness for short, dry turf near the sea is well known, but it also grows on ancient earthworks and limestone pavements. Somewhat bizarrely, it has also developed a liking, too, for people's front lawns. My grandparents have a small population in their garden on the Isle of Wight, nurtured lovingly for the past decade. Under the watchful eye of my grandmother, my grandfather will carefully skirt the area of the lawn where the orchids grow with the lawnmower each summer. They are extremely proud of them. I'd started monitoring this little colony years before first encountering them on the Dorset coast.

Durlston was bathed in a rich golden sunlight as I pulled into the car park in the late afternoon. I had read on their website that the Autumn Lady's-tresses were in flower on the clifftop. This was it. A moment that would hold so much significance for me: it would represent the culmination of my ten-year obsession with seeing every British and Irish orchid. I had come full circle. And so for the fifty-second time, I packed my camera in its bag and swung it over my shoulder, enjoying the eager anticipation that had preceded every species I had seen that year.

The clifftop path was warm. A light breeze played across my bare arms, browned by hours spent orchid hunting in the sun. I thought back to the beginning of the season and the icy winds that had buffeted me and the Early Spider Orchids: what a difference a season makes. The meadows were now a sweep of golden brown, or already cut short ready for the winter. The sea of yellow cowslips had long gone. I passed the little hollow where I had seen my first Early Purple Orchid back in May and grinned: I couldn't help it.

I walked down into the gulley where the writer of *Durlston's daily diary* had reportedly seen some that very morning. Electric Adonis blues and chocolatey brown arguses chased each other through the long grass, buffeted from time to time by the wind. Bromes and oat-grasses shifted softly with metronomic regularity. Looking east, I could make out the Needles running off the end of the Isle of Wight into a glittering turquoise sea.

I slipped into orchid-hunting mode as I began my descent, sweeping the ground with my eyes: back and forth, back and forth, searching for that tiny white spiral that would complete my summer's endeavour. The sward was quite long here, which meant spotting small flowers was harder than usual. Butterflies descended on the purple button-like flowers of common knapweed. Where the grass was shorter there were the bright-pink stars of common centaury and pale-lilac harebells, heads bowed in silent meditation.

In the end it happened very quickly. I was walking down the precipitous side of the gulley when I saw a slender twirl of white. There, poking out from behind a tuft of yellowing grass, was an Autumn Lady's-tresses. Its snowy flowers twisted skyward: a bride on her wedding day. Extraordinary. I sat down, a smile spreading across my face. It was finished. I had seen them all: every last one of them.

I suddenly felt very tired. An adventure that had lasted five months – and many more of planning before that – had come to an end. At the age of twelve, I remember glancing through the few pages at the end of my Collins Wildflower Guide that were decorated with orchids. They were beautiful, rare, mysterious, fanciful and full of intrigue: I longed to find them.

Never did I imagine that I would see all of them. As far as I knew I was the first person to have done so in one summer. It was a humble achievement but one I felt incredibly proud of. During my quest, I had driven nearly 10,000 miles, taken more than 50,000 photos, passed through forty-eight counties and been through two cars. I'd travelled by plane, train, bike and car and had walked miles in search of some of our islands' most precious flowers. It had brought inspiration and ecstasy, panic and fear, happiness, heartbreak and pure, unadulterated Orchidelirium. It had been quite an adventure and, if nothing else, a wonderful way to see the country.

Rushing around Britain had meant spending considerably longer on the road than in the wild confines of nature. Yet every time I arrived in a quiet pocket of the countryside and turned myself over to orchid hunting, I got lost in a world where material concerns became futile in the presence of wildflowers, birdsong and summer sun. Searching for orchids places me in my element, brings me closer to who I really am and allows the stress and strain of everyday life to melt into nothing.

As the early-evening sun flooded the meadows, Lady's-tresses twinkled pearly white in the grass, tiny ivory helices delicately decorating the downland. Down below, the sea crashed rhythmically against the cliffs. As I reached the end of the reserve and turned to walk back, I stood for a moment, looking out over Durlston and the sparkling water, to the Isle of Wight beyond. Taking all of this in, I felt a deep sense of completeness, at having tied up my fascination with these charismatic plants in such fitting circumstances. It had been an exhausting summer, but one I would remember for the rest of my life.

Released from an exhilarating summer of orchid hunting, my thoughts began to turn towards Oxford and beginning the next chapter of my life at university. Would my new friends understand why I had spent my gap year looking for orchids, instead of jetting off to South East Asia? I wondered if I would find people who shared my interests; whether there would be another Suzie. Had that been my one opportunity? Should I have fought harder to make sure she didn't disappear from my life?

Despite completing my task, I didn't feel quite finished with Autumn Lady's-tresses. Two weeks later, I was in the New Forest, on the flat, open plain where I had reached the halfway mark with Lesser Butterfly Orchids earlier in the year – a lone figure, lying on the ground, twisted and contorted as I photographed two Autumn Lady's-tresses side by side. This large expanse of nibbled grass was one of the last places you would expect to find orchids, but there are thousands of Autumn Lady's-tresses here between the round bobbles of horse dung. They were already producing leaves ready for the following year: unlike many orchids, they overwinter as a leaf rosette. When the plant produces a flowering spike, it comes up adjacent to next year's leaves. This was, I supposed, their way of hibernating. I was struck by their remarkable resilience. These tiny, delicate spikes are grazed all summer, trampled underfoot by both humans and ponies, and experience searing summer temperatures out in the open. And yet every autumn they prepare next year's leaves, ready to go all over again.

But as impressive as this is, the world might become a bit too much for them – and the other fifty-one orchids across the country. Climate change is having a strange effect on the orchids in Britain: there are winners and losers. Some – the winners – are benefiting from warmer temperatures and beginning to spread northwards, expanding their range. The Lady Orchid is one of these: historically confined to Kent, it has begun appearing across the south of the country. The Lizard Orchid is also on the increase, with sites being discovered as far west as Bristol in recent years. However, this effect is temporary. As climate change continues along its current trajectory, the south will become too hot and dry for these species and they will be forced to migrate northwards. At the same time, there is likely to be a shift in the composition of the British orchid flora.

As we lose some species, others will be welcomed in from the continent. Small-flowered Tongue Orchids have already been found, whether naturally occurring or not, along the south coast. In 2014, a Sawfly Orchid was found in Dorset, a relative of the sexually deceptive Bee, Fly and Spider Orchids usually found in the Mediterranean basin. In June 2017, as I write, the largest population of Greater Tongue Orchids ever seen in Britain has been discovered in Essex. This exciting new mix of species will be moving north with climate change and, provided they can make the jump across the Channel, the number of records will surely increase over the coming decades.

It won't be a surprise, then, to learn that the losers are more typically northern species like Bog, Coralroot and Small White Orchids. Researchers at Kew Gardens have also found that the Fly Orchid and, surprisingly, the Man Orchid are struggling. Given the Man Orchid is closely related to the Lady Orchid and generally a southern species, we might have expected it to be temporarily benefiting from increasing temperatures. Such puzzles highlight the importance of conducting long-term studies to establish the effects of a warming climate.

In the 1970s, Professor Mike Hutchings of the University of Sussex started what became the longest-running ecological study of any orchid species. For thirty-two years, he and his colleagues meticulously mapped and recorded a population of Early Spider Orchids on the South Downs near Brighton. Each spring, they would take detailed measurements of size, whether the orchids flowered or set seed, when they flowered and how many blooms were produced. One study with this vast dataset, led by Karen Robbirt, showed that the flowering period for these orchids changes from year to year, as you might expect. In 2013, they would have flowered much later than normal, because of the late arrival of spring. While it is impossible to extract any meaningful pattern over a few years, this thirty-two-year dataset has been used to show that for every one-degree increase in mean spring temperature, the orchids start to flower six days earlier. The same pattern is observed when analysing herbarium specimens – plants collected, dried and pressed – in conjunction with historical weather records from the past 167 years. These studies have revealed how flowering is affected by weather and a changing climate.

Early Spider Orchids are clearly being affected by a warming climate, but an even bigger danger could soon present itself. It turns out that the emergence period of *Andrena* bees is much more susceptible to climate change than the flowering period of the orchids. The female bees will be emerging much earlier, coinciding with the orchids flowering. If presented with a choice, the male bees will almost always choose the real female, rather than the fake. So, as the climate warms, the relationship between insect and plant will be knocked out of synchrony. We could be seeing a new kind of calamity for the orchids that has nothing to do with the threat of over-zealous Victorian orchid hunters.

True, despite the fervour of Victorian collectors and the threats of climate change, Britain has, to date, only actually lost one orchid to extinction: the Summer Lady's-tresses. Given our natural bias towards saving good-looking organisms, it is hardly surprising that orchids have been a focus point for many conservationists in the UK (I do wonder whether the same level of protection that's provided for the Lady's Slipper would have been given to the Narrow-lipped Helleborine were it in the same perilous position). Nonetheless, despite the best efforts of Natural England, Plantlife, the Wildlife Trusts and countless other organisations, almost half of our native orchids are classified as Red List species, meaning they are threatened in the UK. Worryingly, some of these species are often thought to be relatively common, like Autumn Lady's-tresses and Greater Butterfly Orchid.

So why put such an emphasis on protecting orchids? Our native species, though beautiful, are not particularly useful. They don't, as far as we know, hold the key to curing illnesses or providing food. Firstly, conserving orchids has a positive impact on other, less threatened species. They are extremely fussy plants, meaning that rarity often reflects very specific growing conditions. The intricacies of orchids are also an exciting way in which to get children interested in the natural world. A flower that looks like a bee? That was enough for me.

But while many of our rarest orchids are available to visit in nature reserves, they often lack a certain wildness that would serve as inspiration for children. When I was a teenager, I hated going to organised reserves, preferring to plunge into nature where it was wild and uncontrolled. The rarest orchids are almost always in nature reserves and in some cases they are put behind bars. Military Orchids in the Chilterns and Late Spider Orchids at Wye are caged, while the Monkey Orchids at Hartslock and Lady's Slippers at Gait Barrows are roped off as if they're in a botanical zoo. Obviously this is a necessary requirement to protect the plants from grazers and human trampling – and we would certainly be in a sorry state without them – but throughout the summer I lamented that it has reached this stage. During my childhood, I spent hours fantasising about coming across populations of rare orchids in the depths of the countryside. Discovering a forgotten glade in the woods, scattered with the royal purple and lilac spikes of Military Orchid is every orchidophile's dream.

Rarity, beauty and wildness should surely be intrinsically linked, epitomising that magical, mysterious power that nature can hold over you. But it is not always the case. With structured nature reserves, the thrill of the chase is, at least to some extent, plucked from your grasp, the rarities served to you on a plate. The species I enjoyed finding the most that summer were the ones holed up in remote, wild corners of the country: the Frog Orchids on the Outer Hebrides, for example, or the unassuming Dense-flowered Orchid that took so long to find in the Burren. The feeling of satisfaction and relief after hours of stressful searching for Fen Orchids and Lindisfarne Helleborines was second to none. These were places where you could dive in and not worry about paths, or where you were and weren't allowed to go, and how you should feel about different things.

So are nature reserves helping our orchids? The answer, unquestionably, is yes: without them we would undoubtedly have lost more than just Summer Lady's-tresses. But by investing so much in protecting individual species, they risk becoming as separated from the wild as from the forces that are wreaking destruction. Perhaps we also need nature reserves to place more emphasis on guarding wilderness, rather than just particular species or habitats.

This boils down to an age-old debate in conservation: should we protect the individual or the environment that allowed that individual to evolve? This was a question that had to be answered at Hartslock ten years ago. With the population of Lady x Monkey hybrids breeding rampantly, the Monkey Orchid gene pool is at risk of dilution. Given this is one of three Monkey Orchid populations in the country, many people were understandably concerned and some would have seen the hybrids removed from the hillside. Those involved with managing the site have decided otherwise, though, opting to protect process rather than individuals.

This is a debate highly relevant to my current line of work. Having enjoyed a wonderful three years studying biology at university, I have somewhat inevitably returned to orchids. My PhD is looking at the genus *Orchis*, and specifically the four British anthropomorphic species: Man, Lady, Monkey and Military Orchids. These plants look very different. So different, in fact, that anyone could sit down with a simple botanical key and identify each one having never seen or heard of them before. And yet hybridisation is a common phenomenon.

What I saw occurring between Monkey and Lady Orchids at Hartslock was just the beginning. Across Europe, whenever any of these four species grow together, they simply can't keep their hands off one another and hybridise rampantly, forming hordes of intermediate orchids. These hybrid plants are then able to successfully pollinate each other and their parent plants, resulting in extensive hybrid swarms; where individual specimens occupy a varied position on a more or less continuous scale between one parent and the other. Sometimes they look like one of the parents, while others are perfect intermediates. My task, then, is to begin to understand why these four species remain distinct. Why haven't they formed one large hybrid super-species? Why do they continue to trade genes with each other? And what can we do to protect endangered populations of these orchids?

I'm incredibly lucky to have the opportunity to develop my interest in these plants further, and to apply what I discover to conservation efforts across Britain and Europe. It's exciting to be at the forefront of science and helping to protect my beloved orchids. There will inevitably be a few people who don't appreciate my anthropomorphising of the orchids, but so many species are predisposed towards such comparisons. How can you not see a frolicking frog, a mischievous monkey or a bumbling bee when it's presented on a plate for you?

I sat cross-legged on the short turf in the middle of the New Forest, all this ahead of me, though I didn't know it then.

My quest was complete. I had laughed, wept and seen some beautiful places. I had been soaked and stung, roasted and bitten, sunburned, rejected, rendered speechless and struck by the generosity of some incredible people. I'd learned about genetics, evolution, climate change and the stranger margins of British plant-hunting history.

Orchids had been at the centre of my life for more than a decade. From that first Bee Orchid on Figsbury Ring, to this summer spent hunting the whole clan, it had been an incredible journey.

A blackbird, tender and melodious, began serenading the open plain as the day drew to a close. The air was still and laced with the familiar aroma of heather and peat. And as the sun sank towards the arboreal horizon, with 10,000 Autumn Lady's-tresses bathed in the fragile warmth of September, my Orchidelirium was finally laid to rest.

Acknowledgements

My summer in search of British and Irish orchids was the grand finale of a childhood steered by a botanical obsession. While the words, photos and experiences are mine, none of them would be on paper without the help of some wonderful people. Many friends have given me their time, but I would like to start by thanking Steve Povey, Sean Cole, Dom Price, Timothy Jenkins, Mike Waller, Steff Leese, Sharon Parr, David and Frances Farrell, Jeff Hodgson, Michael Powell, Paul Drummond, Colin Auld, Gerry Sharkey, Peter Chapman and Paul Smith, for kindly offering me site information and/ or accompanying me on my hunt for orchids. A big thank you to the Horner family from Winterslow Coachworks for keeping me on the road throughout the summer. Thanks also to my PhD supervisors for their patience and understanding while I finished writing.

For access to letters of correspondence between botanists, I'd like to thank the Druce Herbarium at the University of Oxford, the University of Reading Herbarium, the Orchid Herbarium at the Royal Botanic Gardens, Kew, the University of Cambridge Herbarium (CGE) and the Natural History Museum, London (BM). Particular thanks to John Hunnex (BM) and Christine Bartram (CGE) for their time and effort in helping me. I'm also very grateful for the bursaries and grants awarded by the Royal Horticultural Society, the Glasgow Natural History Society, Bishop Wordsworth's School, Charterhouse, and the Sarum St Michael Educational Charity.

To everyone who read through drafts and helped me find a publisher, thank you for believing that it would happen, and thanks to everyone at Short Books who made it happen. In particular, I want to thank my brilliant editor Aurea Carpenter, for taking my text and running with it, and Evie Dunne for her beautiful illustrations.

Finally, my heartfelt thanks go to the following: Cody Sands, Nikki Webber, Alice Thomson and the rest of my friends (you know who you are), for your love, support and encouragement in the face of my rambling, whimpering and relentless requests for opinions. To Suzie Lane, for the hours we spent orchid hunting together (let me know when you find a Ghost). To Betsy Tobin, for your generous offering of time and literary expertise, and for guiding me through the publishing process from start to finish. To Mattie Brindle, without whom I would never have finished writing my story: thank you for helping me see I could still do it. To my best friend, Ben Ingledow: thanks for bouncing back opinions, for picking up the pieces when it all got a bit much, and for constantly cheering me on even as I crawled over the finish line. Lastly, thank you to my family, for twenty-three years of love, support, and trips to look for orchids. I'm sorry if you're embarrassed by any of it.

You're all brilliant, and I owe you so much.

Leif

August 2017

Bibliography

Allen, D. E. (1986) *The Botanists*. St Paul's Bibliographies.

Baker, M. (2008) *Discovering the Folklore of Plants*. Shire Publications Ltd.

Bateman, R. M. (2006) 'How many orchid species are currently native to the British Isles?' *Current Taxonomic Research on the British and European Flora*. Botanical Society of Britain & Ireland.

Bateman, R. M. et al. (2008) 'Morphometric and population genetic analyses elucidate the origin, evolutionary significance and conservation implications of *Orchis* x *angusticruris (O. purpurea* x *O. simia)*, a hybrid orchid new to Britain'. *Botantical Journal of the Linnaean Society* Vol 157: 687-711.

Brooke, J. (1948) *The Military Orchid*. The Bodley Head.

Brooke, J. (1950) *The Wild Orchids of Britain*. The Bodley Head.

Clark, M. (2009) 'Orchids of Kenfig NNR, South Wales'. *Journal of the Hardy Orchid Society* Vol. 6 No. 2: 52.

Clarke, W. A. (1900) *First Records of British Flowering Plants*. West, Newman & Co.

Cole, S. R. (2014) 'History and status of the Ghost Orchid *(Epipogium aphyllum, Orchidaceae)* in England'. *New Journal of Botany* Vol. 4 No. 1.

Culpeper, N. (1652) *The English Physitian.*

Darwin, C. (1892) *The Various Contrivances by which British and Foreign Orchids are Fertilised by Insects, and on the Good Effects of Intercrossing.* John Murray.

Dony, J. G. (1977) 'J. Edward Lousley (1907–1976)'. *Watsonia* Vol 11: 282-286

Endersby, J. (2016) *Orchid: A Cultural History.* University of Chicago Press and Kew Publishing.

Foley, M. & Clarke, S. (2005). *Orchids of the British Isles.* Griffin Press Publishing Ltd.

Gasson, M. (2013) 'Pollination in the Early Purple Orchid'. *Journal of the Hardy Orchid Society* Vol. 10 No. 2.

Gerard, J. (1597) *The Herball, or, Generall historie of plantes.* John Norton

Greene, E. (1562) *Landmarks of Botanical History; a Study of Certain Epochs in the Development of the Science of Botany.* Smithsonian Institution.

Gribbin, M. & Gribbin, J. (2008) *Flower Hunters.* Oxford University Press.

Grigson, G. (1975) *The Englishman's Flora.* Helicon.

Hanbury, F. J. & Marshall, E. S. (1899) *Flora of Kent.* Frederick J. Hanbury.

Harrap, S. & Harrap, A. (2005) *Orchids of Britain & Ireland.* 2nd Edition. A&C Black Publishers Ltd.

Hibbert, A. (2009) *The Long Haul.* Tricorn Books.

Jefferies, R. (1879) *Wild Life in a Southern County.* Smith, Elder & Co.

Bibliography

Lang, D. C. (1980) *Orchids of Britain and Ireland*. Fakenham Press Ltd.

Lee, J. (2015) *Yorkshire Dales*. William Collins Books.

Lousley, J. E. (1969) *Wild Flowers of Chalk & Limestone*. William Collins Books.

Mabey, R. (2010) *A Brush with Nature*. BBC Books.

Nelson, E.C. & Walsh, W. (1997) *The Burren: A Companion to the Wildflowers of an Irish Limestone Wilderness*. Samton Ltd.

Orlean, S. (1998) *The Orchid Thief*. Vintage.

Praeger, R.L. (1909) *A Tourist's Flora of the West of Ireland*. Hodges, Figgis and Co.

Pratt, A. (1873) *The Flowering Plants, Grasses, Sedges, and Ferns of Great Britain, and their Allies the Club Mosses, Pepperworts, and Horsetails*. Frederick Warne and Co.

Rose, F. (2006) *The Wild Flower Key* (Clare O'Reilly Revised Edition) – *How to identify wild plants, trees and shrubs in Britain and Ireland*. Warne

Salisbury, E. (1952) *Downs & Dunes: their Plant Life and its Environment*. G. Bell & Sons Ltd.

Step, E. (1905) *Wild Flowers Month by Month in their Natural Haunts*. Frederick Warne & Co. Ltd.

Summerhayes, V. S. (1951) *Wild Orchids of Britain*. Collins Clear-type Press.

Sumpter, J. P. (2004) 'The Current Status of Military *(Orchis militaris)* and Monkey *(Orchis simia)* Orchids in the Chilterns.' *Watsonia* Vol 25: 175-183

Tahourdin, C. B. (1925) *Native Orchids of Britain*. H. R. Grubb.

Taylor, L. & Roberts, D. L. (2011) 'Biological Flora of the British Isles: *Epipogium aphyllum*' Sw. *Journal of Ecology* Vol 99: 878–890

Townsend, F. (1883) *Flora of Hampshire*. J. Reeve & Co.

Turner Ettlinger, D. M. (1976) *British & Irish Orchids: A Field Guide*. The Macmillan Press Ltd.

Vickery, R. (2010) *Garlands, Conkers and Mother-die: British and Irish Plant-lore*. Continuum.

Webster, A. D. (1898) *British Orchids*. J. S. Virtue & Co.

Letters about the Ghost Orchid from Cambridge University Herbarium; Druce Herbarium, Oxford; Natural History Museum, London; Royal Botanic Gardens, Kew; and University of Reading Herbarium

Species of British Orchid

Listed in alphabetical order.
Page numbers denote first significant
appearance of orchid in the text.

Autumn Lady's-tresses, *Spiranthes spiralis*   337

Bee Orchid, *Ophrys apifera*   229

Bird's-nest Orchid, *Neottia nidus-avis*   110

Bog Orchid, *Hammarbya paludosa*   282

Broad-leaved Helleborine, *Epipactis helleborine*   315

Burnt Orchid, *Neotinea ustulata*   144

Chalk Fragrant Orchid, *Gymnadenia conopsea*   201

Common Spotted Orchid, *Dactylorhiza fuchsii*   67

Common Twayblade, *Neottia ovata*   83

Coralroot Orchid, *Corallorhiza trifida*   185

Creeping Lady's-tresses, *Goodyera repens*   302

Dark Red Helleborine, *Epipactis atrorubens*   279

Dense-flowered Orchid, *Neotinea maculata*   94

Dune Helleborine, *Epipactis dunensis*   291

Early Marsh Orchid, *Dactylorhiza incarnata*   92

THE ORCHID HUNTER

Early Purple Orchid, *Orchis mascula* 30

Early Spider Orchid, *Ophrys sphegodes* 34

Fen Orchid, *Liparis loeselii* 218

Fly Orchid, *Ophrys insectifera* 100

Frog Orchid, *Dactylorhiza viridis* 266

Ghost Orchid, *Epipogium aphyllum* 307

Greater Butterfly Orchid, *Platanthera chlorantha* 149

Green-flowered Helleborine, *Epipactis phyllanthes* 293

Green-winged Orchid, *Anacamptis morio* 36

Heath Fragrant Orchid, *Gymnadenia borealis* 207

Heath Spotted Orchid, *Dactylorhiza maculata* 206

Hebridean Marsh Orchid, *Dactylorhiza
  traunsteinerioides* ssp. *francis-drucei* var. *ebudensis* 265

Irish Lady's-tresses, *Spiranthes romanzoffiana* 323

Irish Marsh Orchid, *Dactylorhiza kerryensis* 56

Lady Orchid, *Orchis purpurea* 81

Lady's Slipper, *Cypripedium calceolus* 168

Late Spider Orchid, *Ophrys fuciflora* 243

Lesser Butterfly Orchid, *Platanthera bifolia* 205

Lesser Twayblade, *Neottia cordata* 192

Lindisfarne Helleborine, *Epipactis sancta* 299

Lizard Orchid, *Himantoglossum hircinum* 238

Loose-flowered Orchid, *Anacamptis laxiflora* 65

Species of British Orchid

Man Orchid, *Orchis anthropophora* 75

Marsh Fragrant Orchid, *Gymnadenia densiflora* 294

Marsh Helleborine, *Epipactis palustris* 220

Miltary Orchid, *Orchis militaris* 137

Monkey Orchid, *Orchis simia* 129

Musk Orchid, *Herminium monorchis* 273

Narrow-lipped Helleborine, *Epipactis leptochila* 314

Northern Marsh Orchid, *Dactylorhiza purpurella* 258

Pugsley's Marsh Orchid, *Dactylorhiza
    traunsteinerioides* 97

Pyramidal Orchid, *Anacamptis pyramidalis* 216

Red Helleborine, *Cephalanthera rubra* 253

Small White Orchid, *Pseudorchis albida* 214

Southern Marsh Orchid, *Dactylorhiza praetermissa* 66

Sword-leaved Helleborine, *Cephalanthera longifolia* 108

Violet Helleborine, *Epipactis purpurata* 333

White Helleborine, *Cephalanthera damasonium* 126